167268

PERSPECTIVES ON OPTIMIZATION

PERSPECTIVES ON OPTIMIZATION

A Collection of Expository Articles

Edited by
A. M. GEOFFRION
University of California, Los Angeles

♠ ADDISON-WESLEY PUBLISHING COMPANY

Reading, Massachusetts · Menlo Park, California · London · Don Mills, Ontario

PREFACE

Most instructors in a fast-moving field like optimization find it valuable or even essential to supplement their chosen textbook by selected readings from the recent literature. Expository/survey articles are particularly useful when they offer simplifying insights, unifying principles or sympathetic communication of the genuine essence of some significant portion of the field. It is disappointing that there has always been a scarcity of such articles. The situation is further aggravated by the expense and nuisance of multiple reproduction from widely scattered sources for individual student use. The present volume is intended to help alleviate these difficulties, and hopefully also to encourage other instructors to make similarly available their own favorite eclectic collections of materials for the student.

This book contains a selection of eight critical surveys on various aspects of optimization. The primary (but not exclusive) emphasis is on optimization in a static rather than dynamic or control-theoretic setting, and on useful theory and computationally effective algorithms rather than applications *per se*. Five of the contributions have been recently published in journal form, two others will be shortly, and the eighth is based on portions of a forthcoming book (the title page of each chapter gives the exact reference). Nearly all of the chapters should be readable by students with undergraduate preparation in mathematics, and with a modest prior background in optimization.

The book is organized into four parts. Part I, on unconstrained optimization, consists of M. J. D. Powell's authoritative account of techniques available up to 1968. Interested readers may also wish to consult his sequel survey covering subsequent contributions, "Recent Advances in Unconstrained Optimization," forthcoming in the new journal *Mathematical Proramming*, Vol. 1, No. 1.

Part II is on constrained optimization with continuous-valued variables. In Chapter 2, I attempt to identify the fundamental concepts and explain the main computational approaches for large-scale mathematical programming. In Chapter 3, I develop duality and optimality theory for nonlinear programming in a way that seems to facilitate its successful application. Chapter 4, by D. C. Luenberger, freshly illuminates the evolutional relationships between the sister fields of mathematical programming and optimal control theory.

Part III is devoted to the very active area of constrained optimization with integer-valued variables. R. E. Marsten and I present, in Chapter 5, a unified viewpoint of the four main general computational approaches to integer programming. Extensive computational experience is cited. As counterpoint to this general discussion, and to stress that the exploitation of special problem structure can be very fruitful, Chapter 6 by R. Garfinkel and G. L. Nemhauser is devoted to one of the most important special problem classes: the set covering problem.

Part IV treats optimization in graphs and networks. Chapter 7 is a broad elementary exposition by the principal founder and contributor in this field, D. R. Fulkerson. The final chapter is an incisive analysis by S. E. Dreyfus of algorithms for the shortest-path problem. This class of problems is of major importance both in its own right and as a key subproblem in algorithms for much more complex problems involving networks. Professor Dreyfus has also kindly provided a new postscript to his original article.

The primary intended use of this book is as a supplementary text for a variety of different types of courses on optimization: linear programming, non-linear programming, integer or combinatorial programming, large-scale programming, network flows, and others. Students taking several courses on optimization should eventually digest all of the chapters, though perhaps only a few of them in connection with any one course. Primary intent notwithstanding, however, several of the broader chapters are in fact comprehensive enough to serve as the backbone of an entire course, and have been so used.

It is a pleasure to express my appreciation to all of the contributors to this book, and to the copyright holders who granted permission for republication.

Los Angeles, California   A. M. G.
October, 1971

ACKNOWLEDGMENTS

The editor of this collection wishes to thank the following individuals and organizations for permission to reprint the material listed below:

Society for Industrial and Applied Mathematics

M. J. D. Powell, "A Survey of Numerical Methods for Unconstrained Optimization," *SIAM Review,* **12,** 1 (January 1970), 79–97.

A. M. Geoffrion, "Duality in Nonlinear Programming: A Simplified Applications-Oriented Development," *SIAM Review,* **13,** 1 (January 1971), 1–37.

The Institute of Management Sciences

A. M. Geoffrion, "Elements of Large-Scale Mathematical Programming," *Management Science,* **16,** 11 (July 1970), 652–691.

A. M. Geoffrion and R. E. Marsten, "Integer Programming Algorithms: A Framework and State-of-the-Art Survey," *Management Science,* **18,** 7 (March 1972).

The Institute of Electrical and Electronics Engineers, Inc.

D. G. Luenberger, "Mathematical Programming and Control Theory: Trends of Interplay," *IEEE Transactions on Automatic Control,* in press.

Wiley-Interscience, A Division of John Wiley & Sons, Inc. Publishers

R. Garfinkel and G. L. Nemhauser, "Optimal Set Covering: A Survey," based on a chapter of their forthcoming book *Integer Programming,* Wiley, 1972, and also on the unpublished paper "Set Covering: A Survey," presented by R. Garfinkel at the 17th International Conference of the Institute of Management Sciences, July 1970.

D. R. Fulkerson

"Flow Networks and Combinatorial Operations Research," originally published in *American Mathematical Monthly,* **73,** 2 (February 1966), 115–138.

Operations Research

S. E. Dreyfus, "An Appraisal of Some Shortest-Path Algorithms," *Operations Research,* **17,** 3 (May-June 1969), 395–412.

CONTENTS

I. UNCONSTRAINED OPTIMIZATION

1. A SURVEY OF NUMERICAL METHODS FOR UNCONSTRAINED OPTIMIZATION*

M. J. D. POWELL†

Abstract. This paper is intended to introduce and relate the more successful numerical methods for calculating the greatest value of a given real function $F(x_1, x_2, \ldots, x_n)$. It also indicates some directions for research.

1. Introduction. We consider the problem of calculating the greatest value of a given real function $F(x_1, x_2, \cdots, x_n)$, where each variable x_i, $i = 1, 2, \cdots, n$, can take the value of any real number. We wish to carry out the calculation on a computer, so the objective function is usually defined by a subroutine that will evaluate $F(x_1, x_2, \cdots, x_n)$ for any choice of the vector of variables. Therefore the problem is solved by applying a systematic method to try many different values of the variables, the choice of variables being guided by the values of the objective function that have already been calculated. Some methods for adjusting the variables use strategies that depend on derivatives of $F(x_1, x_2, \cdots, x_n)$, in which case the computer subroutine that specifies the objective function has to be extended to calculate derivatives as well.

Because so many different algorithms have been suggested to solve the unconstrained optimization problem, this paper is intended to introduce the subject in a concise and uncomplicated way. Therefore we deliberately omit many details that are needed to write computer programs, because we try to emphasize the ideas that cause the success of the algorithms, rather than the techniques themselves. In particular we dwell on the ideas either that have proved to be basic to the subject, or that, in the author's opinion, are likely to yield methods that are more successful than the ones of today. Thus we indicate some areas of research, and here we must emphasize that at the present time the chances of finding substantial improvements to existing algorithms seem to be excellent.

Other papers that relate different methods of optimization have been written by Box [5], Dickinson [11], Fletcher [13], Leon [24], Powell [30] and Spang [34], and a more detailed introduction to the subject will soon be published, written by Box, Davies and Swann [6].

2. The objective function. In order to calculate the greatest value of the objective function, $F(x_1, x_2, \cdots, x_n)$, by a general algorithm, it is necessary for $F(x_1, x_2, \cdots, x_n)$ to satisfy some smoothness conditions. It is not possible to be specific, but it is appropriate to make some general remarks here, because we must indicate that we cannot expect any algorithm to be entirely successful.

The reason is that, because each component of the vector of variables $(x_1, x_2, \cdots, x_n)$ is free to take the value of any real number, we cannot begin to cover

* Received by the editors July 23, 1968. Presented by invitation at the Symposium on Optimization, sponsored by the Air Force Office of Scientific Research, at the 1968 National Meeting of Society for Industrial and Applied Mathematics, held in Toronto, Canada, June 11–14, 1968.

† Mathematics Branch, Theoretical Physics Division, Atomic Energy Research Establishment, Harwell, Berkshire, England.

the whole space of the variables. Further, even if from the structure of $F(x_1, x_2, \cdots, x_n)$ we know that the required maximum must lie in a finite region, say, $0 \leq x_i \leq 1$, $i = 1, 2, \cdots, n$, then often a covering is not feasible even for quite small values of n, because a square mesh of size h in the unit cube demands $(1 + h^{-1})^n$ points. Therefore, a realistic ambition is to develop a general algorithm whose failure will be infrequent when it is applied to problems of the real world.

Most algorithms economize on the number of function evaluations by searching near the point at which the largest value of $F(x_1, x_2, \cdots, x_n)$ has been calculated. In the case $n = 2$ it is easy to visualize this idea, for $z = F(x_1, x_2)$ describes a surface in three dimensions, and we require the point of the surface that is furthest above the plane $z = 0$. Also the analogy shows a disadvantage of the general method, which is that if the best calculated value of $F(x_1, x_2, \cdots, x_n)$ happens to be on a small local hill of the surface, then an algorithm is likely to find and to stop at the top of the small hill, even if there are higher hills. Because of the difficulty of searching right through the space of the variables, this deficiency seems to be inevitable in general optimization methods. Therefore, for guaranteed success, it is often necessary for $F(x_1, x_2, \cdots, x_n)$ to have only one maximum point, but unfortunately functions with more than one maximum are often encountered in practice.

If there are many maxima, the one that is found usually depends on an initial estimate of the solution, which has to be provided by the user of the algorithm. In this case different initial estimates can be tried if it is suspected that $F(x_1, x_2, \cdots, x_n)$ is awkward, but this device alone can only reduce, rather than eliminate, the possibility of failure.

For these reasons we judge an algorithm to be successful if it calculates a local maximum of $F(x_1, x_2, \cdots, x_n)$, from a given starting approximation to the required vector of variables. This statement and the remark that optimization can be visualized as a hill-climbing problem indicate the basic conditions that must be satisfied by the objective function. Also some algorithms require $F(x_1, x_2, \cdots, x_n)$ to be differentiable.

The lack of definiteness in stating the conditions on $F(x_1, x_2, \cdots, x_n)$ is deliberate, because we are describing the current state of optimization, and it happens that the current state is not a logical structure of theorems. Instead it has developed from an assortment of numerical methods which have been devised because real problems have had to be solved, and at present the actual success of the algorithms is far ahead of any theoretical predictions. Fortunately there seems to be an increase in the amount of study on theoretical questions, so the subject of unconstrained optimization should soon become more coherent.

3. Old methods. In §§ 3–7 we follow the history of optimization algorithms, and we mention various methods that have been in use for at least three years. However, in §§ 8–10 we consider some recent ideas, and also we suggest some future developments.

The celebrated steepest ascent method was described by Cauchy [8] in 1847. In the case $n = 2$, when the objective function is the surface $z = F(x_1, x_2)$, the method is obtained by asking the question, "Which uphill direction is steepest?" If the objective function is differentiable, and we take a very small step (δ_1, δ_2) from the point (x_1, x_2), we obtain from first order terms in the Taylor series the result:

(1)

$$F(x_1 + \delta_1, x_2 + \delta_2) - F(x_1, x_2) \approx \delta_1 \frac{\partial F}{\partial x_1} + \delta_2 \frac{\partial F}{\partial x_2}$$

$$= \sqrt{\delta_1^2 + \delta_2^2} \sqrt{\left(\frac{\partial F}{\partial x_1}\right)^2 + \left(\frac{\partial F}{\partial x_2}\right)^2} \cos \theta,$$

where θ is the angle between the step (δ_1, δ_2) and the gradient vector at (x_1, x_2). Different directions of the step give different angles, and (1) suggests that the steepest direction occurs when $\cos \theta$ is equal to one. Therefore a step of the steepest ascent method from the point $(x_1, x_2, \cdots, x_n)$ is chosen to be in the direction $(g_1, g_2, \cdots, g_n)$, where g_i is defined by the equation:

(2) $$g_i(x_1, x_2, \cdots, x_n) = \frac{\partial F(x_1, x_2, \cdots, x_n)}{\partial x_i}, \qquad i = 1, 2, \cdots, n.$$

The length of the step is often calculated to maximize the new value of the objective function, in which case an iteration replaces an estimate $(x_1, x_2, \cdots, x_n)$ of the position of the optimum by the estimate $(x_1 + \lambda^* g_1, x_2 + \lambda^* g_2, \cdots, x_n + \lambda^* g_n)$, where λ^* is the value of λ that maximizes the function of one variable:

(3) $$\phi(\lambda) = F(x_1 + \lambda g_1, x_2 + \lambda g_2, \cdots, x_n + \lambda g_n).$$

Many other optimization algorithms are like the steepest ascent method, for they calculate a direction at an estimate $(x_1, x_2, \cdots, x_n)$, and they change the estimate by a multiple of the direction, the multiplier being chosen to maximize the new value of $F(x_1, x_2, \cdots, x_n)$. Therefore the problem of calculating the maximum value of a function of one variable, like expression (3), often needs to be solved. We prefer not to describe the methods of solution here (techniques are given in references [6], [14], [29] and [34]), because most of the details do not contribute to the understanding of many variable problems. However, there is one detail that should be mentioned, which is that it is usually inefficient to seek high accuracy, when the one-dimensional problem is derived from a poor estimate $(x_1, x_2, \cdots, x_n)$ of the solution of the n-dimensional optimization. Box [5] makes some interesting remarks on this question.

The steepest ascent algorithm is one of the few optimization methods that is supported by convergence theorems [20]. If $F(x_1, x_2, \cdots, x_n)$ is bounded above, and has a uniformly continuous first derivative, and if each function $\phi(\lambda)$ (see (3)) has a maximum, then the successive gradient vectors $(g_1, g_2, \cdots, g_n)$ that are generated by the iterations of the algorithm will converge to zero. Therefore any limit point of the sequence of estimates $(x_1, x_2, \cdots, x_n)$ must be a stationary point of $F(x_1, x_2, \cdots, x_n)$. Further, if both the initial estimate $(x_1, x_2, \cdots, x_n)$ and the objective function are such that all the estimates $(x_1, x_2, \cdots, x_n)$ are in a compact space, and also if $F(x_1, x_2, \cdots, x_n)$ has separate stationary points and is not constant along the straight line joining any two of its stationary points (this last condition rules out the possibility that the iterations oscillate between one stationary point and another), then the steepest ascent algorithm will converge to a distinct stationary point of the objective function. The point need not be a maximum, but fortunately it usually happens that convergence to a stationary point that is not a maximum is unstable.

The rate of convergence has been analyzed by Akaike [1], and has been experienced by very many computer users. It is often intolerably slow, so the algorithms described later in this paper are usually more satisfactory. Indeed the poor performance of the steepest ascent method was a strong motivation in the development of recent techniques.

Often it happens that first derivatives of the objective function are not available, either because they do not exist, or, more usually, because one prefers not to calculate them. In this case a classical method of optimization is to adjust each variable separately. Specifically, the initial estimate $(x_1, x_2, \cdots, x_n)$ is altered to $(x_1 + \lambda_1, x_2, \cdots, x_n)$, this estimate is altered to $(x_1 + \lambda_1, x_2 + \lambda_2, x_3, \cdots, x_n)$, and so on until the nth stage of the process replaces $(x_1 + \lambda_1, x_2 + \lambda_2, \cdots, x_{n-1} + \lambda_{n-1}, x_n)$ by $(x_1 + \lambda_1, x_2 + \lambda_2, \cdots, x_n + \lambda_n)$. These n stages are repeated, usually until the corrections $(\lambda_1, \lambda_2, \cdots, \lambda_n)$ become very small. One way of obtaining the values of the numbers λ_i is to use a one-dimensional maximization algorithm to make each new value of the objective function as large as possible.

Because this process can increase $F(x_1, x_2, \cdots, x_n)$ unless all the gradient components $g_i, i = 1, 2, \cdots, n$, are zero, the convergence properties are like those of the steepest ascent algorithm. In particular the rate of convergence is usually too slow to be useful. However, sometimes the structure of $F(x_1, x_2, \cdots, x_n)$ is such that the successive evaluations of the objective function can be speeded up considerably, because the algorithm changes only one of the n variables at each stage.

4. Newton-Raphson. As its name suggests, the "Newton-Raphson" algorithm is also an "old method", but we describe it in a new section because it is still the best way of solving certain problems. Its special feature is that it takes account of second derivatives of $F(x_1, x_2, \cdots, x_n)$. Second derivative terms are absolutely basic to unconstrained maximization, because a differentiable function cannot have a distinct maximum point unless it curves. At this stage our ideas depart from the concepts of linear programming.

The Newton-Raphson method estimates the position of the maximum of $F(x_1, x_2, \cdots, x_n)$ from second and lower order terms in the Taylor series. Specifically it uses the approximation:

$$
\begin{aligned}
& F(x_1 + \delta_1, x_2 + \delta_2, \cdots, x_n + \delta_n) \\
& \approx F(x_1, x_2, \cdots, x_n) + \sum_{i=1}^{n} \delta_i g_i + \tfrac{1}{2} \sum_{i=1}^{n} \sum_{j=1}^{n} \delta_i G_{ij} \delta_j,
\end{aligned}
$$

(4)

where g_i and G_{ij} are defined by (2) and by the equation:

(5)
$$
G_{ij} = \frac{\partial^2 F(x_1, x_2, \cdots, x_n)}{\partial x_i \partial x_j}.
$$

At the maximum of a differentiable function all first derivatives are equal to zero, so, if (4) is exact, the point $(x_1 + \delta_1, x_2 + \delta_2, \cdots, x_n + \delta_n)$ is the required optimum only if the equations

(6)
$$
g_i + \sum_{j=1}^{n} G_{ij} \delta_j = 0, \qquad\qquad i = 1, 2, \cdots, n,
$$

are satisfied. But these equations are linear in the components δ_i, so it is usually straightforward to calculate their solution:

$$(7) \qquad\qquad \delta_i = -\sum_{j=1}^{n} (G^{-1})_{ij} g_j, \qquad\qquad i = 1, 2, \cdots, n.$$

Because of these remarks the Newton-Raphson method changes an estimate $(x_1, x_2, \cdots, x_n)$ of the required vector of variables to the estimate $(x_1 + \delta_1, x_2 + \delta_2, \cdots, x_n + \delta_n)$, where the numbers δ_i are defined by (7), the derivatives G_{ij} and g_j being calculated at the point $(x_1, x_2, \cdots, x_n)$. This technique is applied iteratively, by changing each new estimate in the way described, and it is hoped that the resultant sequence of vectors $(x_1, x_2, \cdots, x_n)$ will converge to the values of the variables that maximize the objective function.

Because the Newton-Raphson method is exact if the approximation (4) holds, it may be proved that, if the second derivative matrix G is negative definite at the solution, then the iterations have quadratic convergence, provided that the initial estimate $(x_1, x_2, \cdots, x_n)$ is sufficiently close to the optimal values of the variables. Therefore the procedure can be very efficient, but, if the initial estimate is poor, the method often fails to converge. Even when it does converge it may reach a point which is not a maximum, because the derivation of an iteration depends only on the fact that the first derivatives of $F(x_1, x_2, \cdots, x_n)$ are zero at a maximum, and the first derivatives are zero at all stationary points of the objective function.

To avoid convergence to a stationary point that is not a maximum, and to ensure that an iteration does not worsen the value of the objective function, a valuable well-known strategy is to use the correction $(\delta_1, \delta_2, \cdots, \delta_n)$, defined by (7), as a direction of search in the space of the variables. (Moreover, (6) can define a direction of search even when G is singular.) Specifically, we treat the function of one variable:

$$(8) \qquad\qquad \phi(\lambda) = F(x_1 + \lambda\delta_1, x_2 + \lambda\delta_2, \cdots, x_n + \lambda\delta_n)$$

in the same way as we treated the function (3). Thus we obtain an algorithm that has both second order convergence, and the ability to converge from very poor starting approximations to the required solution.

This "extended Newton-Raphson" algorithm has been very successful in practice, but there has been little work on identifying general conditions for guaranteed convergence to a local maximum. Powell [30] gives an example to show that the method can fail at a point where the gradient of $F(x_1, x_2, \cdots, x_n)$ is nonzero, and G is nonsingular, so probably any theoretical conditions would exclude many of the successful solutions to real problems that have been obtained already.

It is interesting and instructive to note that the extended Newton-Raphson algorithm is identical to the steepest ascent method, if the second derivative matrix G is equal to minus the unit matrix. It is instructive because, if $-G$ is positive definite (this condition is usually obtained at the solution), and we make the change of variables:

$$(9) \qquad\qquad y_i = \sum_{j=1}^{n} (-G^{1/2})_{ij} x_j,$$

then the second derivative matrix of the right-hand side of (4) becomes equal to minus the unit matrix. For this reason each iteration of the steepest ascent algorithm can be made as successful as a Newton-Raphson iteration by applying a suitable linear transformation to the variables, but the best linear transformation depends on $(x_1, x_2, \cdots, x_n)$. This idea is developed in our discussion of Davidon's method (see § 7).

To many computer users, the most serious disadvantage of the Newton-Raphson algorithm is that it requires second derivatives of the objective function. Therefore techniques have been developed to achieve fast ultimate convergence, by taking the curvature of $F(x_1, x_2, \cdots, x_n)$ into account, without the explicit evaluation of second derivatives. Many of these methods are described later in this paper.

5. Extensions of the old methods. As soon as automatic computers were used to apply the steepest ascent and Newton-Raphson algorithms, it became worthwhile to overcome the limitations of the techniques. In particular attempts were made [4], [12], [17] to find modifications to improve the slow convergence of the steepest ascent method, and to make the Newton-Raphson iteration more reliable.

Booth [4] remarks that it is sometimes worthwhile to modify the steepest ascent method so that it takes shorter steps. Specifically he suggests applying the classical method only every fifth iteration, while on the remaining iterations he replaces the current estimate $(x_1, x_2, \cdots, x_n)$ of the solution by the estimate $(x_1 + 0.9 \, \lambda^* g_1, x_2 + 0.9 \, \lambda^* g_2, \cdots, x_n + 0.9 \, \lambda^* g_n)$, where λ^* is calculated to maximize (3). A good reason for taking these shorter steps is obtained by asking the question: given that we shall apply just two iterations of the steepest ascent method, except that we are free to choose the step-length of the first iteration, what choice of this step-length will cause the final value of $F(x_1, x_2, \cdots, x_n)$ to be greatest? If $F(x_1, x_2, \cdots, x_n)$ is a negative definite quadratic function, it is straightforward to analyze this question, and it happens that, in the usual case when the gradient vector does not point directly at the solution, it is best to choose the initial step to be less than the ordinary step. Presumably, because of the dependence of the steepest ascent method on linear transformations of the variables (see the remarks supporting (9)), the same device can be used to advantage when the direction of search is obtained by multiplying the gradient vector by a constant positive definite matrix. A similar technique is described by Fadeev and Fadeeva [12], and they give an example to show that the shorter steps can provide a faster rate of convergence.

The modification suggested by Forsythe and Motzkin [17] has a more satisfactory mathematical basis. They noticed that the steepest ascent method usually converges in such a way that the asymptotic behavior of the successive estimates $(x_1, x_2, \cdots, x_n)$ is that they alternate between just two lines (in the n-dimensional space of the variables) which meet at the required solution. Some theorems on this behavior have been given by Akaike [1]. The directions of these lines tend to be along the vectors $(x_1^{(k-2)} - x_1^{(k)}, x_2^{(k-2)} - x_2^{(k)}, \cdots, x_n^{(k-2)} - x_n^{(k)})$, where the superscripts are iteration numbers, so the modification introduces occasional searches along these directions. Specifically the estimate $(x_1^{(k+1)}, x_2^{(k+1)}, \cdots, x_n^{(k+1)})$ is calculated usually by searching from $(x_1^{(k)}, x_2^{(k)}, \cdots, x_n^{(k)})$ along the direction

$(g_1^{(k)}, g_2^{(k)}, \cdots, g_n^{(k)})$ for the maximum of $F(x_1, x_2, \cdots, x_n)$, but sometimes the direction of search is $(x_1^{(k-2)} - x_1^{(k)}, x_2^{(k-2)} - x_2^{(k)}, \cdots, x_n^{(k-2)} - x_n^{(k)})$.

A third extension to the steepest ascent algorithm is the "s-step gradient method" [12]. It has been used when the objective function is quadratic, and in this case each iteration is related to s, $1 < s < n$, consecutive iterations of the method of steepest ascents. Specifically an iteration, instead of maximizing the one-dimensional function (3) to provide the best point in the direction $(g_1, g_2, \cdots, g_n)$, calculates the value of $(\lambda_1, \lambda_2, \cdots, \lambda_s)$ that maximizes the function of s variables:

$$(10) \qquad \phi(\lambda_1, \lambda_2, \cdots, \lambda_s) = F\left\{(x_1, x_2, \cdots, x_n) + \sum_{t=1}^{s} \lambda_t(g_{t1}, g_{t2}, \cdots, g_{tn})\right\},$$

where the directions $(g_{t1}, g_{t2}, \cdots, g_{tn})$, $t = 1, 2, \cdots, s$, are the gradients that would be used by s iterations of the steepest ascent method. Using the notation $(\lambda_1^*, \lambda_2^*, \cdots, \lambda_s^*)$ for the calculated vector of parameters, an iteration replaces the estimate $(x_1, x_2, \cdots, x_n)$ by the estimate:

$$(11) \qquad (x_1, x_2, \cdots, x_n) + \sum_{t=1}^{s} \lambda_t^*(g_{t1}, g_{t2}, \cdots, g_{tn}).$$

Forsythe [16] has indicated that the asymptotic behavior of the s-step gradient method is like that of the method of steepest ascents, so we prefer the more recent algorithms that are described in § 7.

An extension to make the Newton-Raphson method more reliable was suggested by Levenberg [25] in 1944, but he applies the idea to the problem of solving overdetermined algebraic equations. A description that is specific to unconstrained optimization is given by Goldfeld, Quandt and Trotter [18], [19]. The extension introduces a nonnegative parameter α that interpolates between the steepest ascent iteration and the Newton-Raphson formula (7), the actual iteration being the replacement of an estimate $(x_1, x_2, \cdots, x_n)$ by the estimate $(x_1 + \delta_1, x_2 + \delta_2, \cdots, x_n + \delta_n)$, where the correction is defined by the equation:

$$(12) \qquad \delta_i = \sum_{j=1}^{n} ([\alpha I - G]^{-1})_{ij} g_j, \qquad\qquad i = 1, 2, \cdots, n,$$

I being the unit matrix. Note that in the case $\alpha = 0$, equation (12) reduces to Newton-Raphson, while if α becomes very large the correction to $(x_1, x_2, \cdots, x_n)$ tends to have the direction of the gradient $(g_1, g_2, \cdots, g_n)$. Moreover, because large values of α cause the length of the correction to be small, there is no need to include a linear search procedure in an iteration. This seems to be one of the most promising ideas for improving existing algorithms, so we shall discuss it further in § 9.

6. Direct search methods. The introduction of automatic computers, as well as stimulating the extensions to classical methods that have just been described, also led to many optimization algorithms that do not require the calculation of derivatives. Most of these methods were guided by extensive numerical experience, and by thinking of the problem as that of climbing a hill, so they developed in an intuitive way. Some very useful techniques resulted, and we mention a few of them now. In the next section other techniques are given that also do not require derivatives to be computed, but they are described separately because, unlike the methods of this section, they originated from an intention to take account of low order terms in a Taylor series expansion of the objective function.

To begin the development of the direct search algorithms, we consider the classical method, described in § 3, of revising the variables of the objective function one at a time. We continue to define an iteration as the process which adjusts each variable once only, and we consider the estimates $(x_1^{(k)}, x_2^{(k)}, \cdots, x_n^{(k)})$, $k = 1, 2, \cdots$, that are obtained by the successive iterations. In hill-climbing terminology, it often happens that this process generates estimates that climb up to the top of a ridge, and then they follow the ridge to the solution, and in fact most of the computer time is usually spent in following the ridge. Rosenbrock noticed that the points on a ridge are often nearly collinear, and he used this remark to provide a successful algorithm [31].

The basis of Rosenbrock's method is that it tries to identify the direction of the ridge in order to use it as a search direction. Initially the variables are changed one at a time, as in the classical method, so on the first iteration the initial estimate $(x_1, x_2, \cdots, x_n)$ is changed to $\{(x_1, x_2, \cdots, x_n) + \lambda_1 \mathbf{d}_1\}$, this estimate is changed to $\{(x_1, x_2, \cdots, x_n) + \lambda_1 \mathbf{d}_1 + \lambda_2 \mathbf{d}_2\}$, and so on, until the complete iteration replaces the initial guess of the solution by the estimate:

$$(13) \qquad (x_1, x_2, \cdots, x_n) + \sum_{t=1}^{n} \lambda_t \mathbf{d}_t,$$

where, for $t = 1, 2, \cdots, n$, $\mathbf{d}_t$ is the tth coordinate direction in the space of the variables. However, before starting a new iteration, the set of n search directions is changed, and the first search direction is replaced by the vector:

$$(14) \qquad \mathbf{d}_1^* = \sum_{t=1}^{n} \lambda_t \mathbf{d}_t,$$

which is the change to the estimate of the solution that has just been calculated. The remaining new search directions are obtained by an orthogonalization process, and then the iterative process is repeated. Thus, if a straight ridge is followed to the solution, the first search direction becomes aligned along the ridge, and so the rate of convergence becomes faster. A further development of this idea is due to Davies, Swann and Campey [37], and some numerical examples are given by Box [5] and Fletcher [13].

Another useful direct search algorithm is described by Hooke and Jeeves [23]. It also seeks to identify the direction of ridges, but it has an important new feature that causes it usually to be even more successful than Rosenbrock's method. This feature comes from the remark that if a long straight step is taken along a curved ridge, then probably the end of the step will fall below the top of the ridge, but this does not matter because it is easy to identify a direction back to the top. Therefore, unlike most other methods, the algorithm of Hooke and Jeeves will sometimes change an estimate $(x_1, x_2, \cdots, x_n)$ of the solution, even though the change causes the value of the objective function to decrease.

An iteration of the Hooke and Jeeves algorithm is in two parts, which are called "pattern move" and "exploratory move". The pattern move is applied first, and it changes the current estimate $(x_1^{(k)}, x_2^{(k)}, \cdots, x_n^{(k)})$ of the position of the solution by the total change made in the last iteration (except that on the first iteration there is no pattern move); let the resultant point be $(y_1, y_2, \cdots, y_n)$. From this point an exploratory move is made, which is really a fine adjustment of the values of the variables. Specifically small steps are taken along each of the coordinate directions in order to increase the objective function; let the resultant point be $(z_1, z_2, \cdots, z_n)$. If $F(z_1, z_2, \cdots, z_n)$ is greater than $F(x_1^{(k)}, x_2^{(k)}, \cdots, x_n^{(k)})$, then $(z_1, z_2, \cdots, z_n)$ becomes the starting approximation for the next iteration, but otherwise the iteration is treated as a failure, and instead an exploratory move is made from the point $(x_1^{(k)}, x_2^{(k)}, \cdots, x_n^{(k)})$.

It should be clear that the pattern moves can take long steps along ridges, while the exploratory phase can climb back to the crest of a ridge. The fact that $F(y_1, y_2, \cdots, y_n)$ is permitted to be less than $F(x_1, x_2, \cdots, x_n)$ is the special feature that causes Hooke and Jeeves method to be particularly suitable for optimizing objective functions that have curved ridges.

The method due to Baer [2] also includes the ideas that are important to Hooke and Jeeves algorithm. The main difference is that Baer uses first derivatives of the objective function, so his exploratory move is a search in the steepest ascent direction. Leon [24] relates these direct search methods to other algorithms.

Another valuable addition to the collection of optimization algorithms was made by Spendley, Hext and Himsworth [35]. It is called the "simplex method", and is quite different from the linear programming technique having the same name. A special property of the simplex method for unconstrained optimization is that its progress depends just on whether certain values of $F(x_1, x_2, \cdots, x_n)$ are greater than or less than other values of the objective function, so it is insensitive to inaccurate function values. Therefore it is a suitable technique when substantial random errors are present in the method for obtaining values of $F(x_1, x_2, \cdots, x_n)$. For instance, if the output of an industrial process depends on time and on certain adjustable input parameters, and if it is required to control these parameters so that the output remains near its optimal level, then the simplex method can be used to regulate the input variables, according to measurements of the objective function.

Its name is derived from the fact that it uses simplexes, a simplex in n-dimensional space being a solid having plane faces and $n + 1$ vertices. For example, triangles and tetrahedra are simplexes in two and three dimensions. The algorithm starts with a given simplex in the space of the variables, and then an iterative procedure is applied to move the given simplex to a position that is near to the optimal point in the space of the variables. Specifically each iteration moves the given simplex by reflecting it in one of its faces, so the position of only one vertex is changed by an iteration. The vertex that is moved is usually the one at which the value of the objective function is least, but there are exceptions to this rule to prevent premature cycling. Once the simplex is close to the optimal point, it can be contracted, in order to determine the required vector of variables more precisely.

Nelder and Mead [27] have extended the simplex method to allow reflection, or expansion, or contraction of the simplex on every iteration. Thus they obtain a procedure that appears to compare favorably with other good algorithms when the objective function is exact and differentiable. However, experiments by Box [5] on objective functions having five and more variables indicate that Nelder and Mead's method is a good algorithm for optimizing an exact function only if the number of variables is small.

7. Conjugate direction methods. We remarked in § 4 that if the first few terms of a Taylor series expansion of the objective function are to be used to predict the solution of the unconstrained optimization problem, then second derivatives must be taken into account. Thus we obtained the Newton-Raphson algorithm, which requires the explicit evaluation of first and second derivatives of the objective function, and we stated that this method has second order convergence. Because this fast convergence is useful if very accurate answers are required, and because the task of calculating second derivatives can be laborious, various algorithms have been proposed that try to take account of second derivative terms, using only values and first derivatives of $F(x_1, x_2, \cdots, x_n)$. Some are described in this section, and we shall find that they use strategies that would yield the exact answer if $F(x_1, x_2, \cdots, x_n)$ were a quadratic function. This property alone does not guarantee fast convergence when the higher derivatives of $F(x_1, x_2, \cdots, x_n)$ are nonzero, but in practice the algorithms are extremely successful, although theoretical reasons for the success have not yet been found. Moreover, numerical examples show that often the methods of this section are the best available, even when the user's initial estimate of the vector of variables $(x_1, x_2, \cdots, x_n)$ is very far from the position of the optimum.

The methods calculate the exact maximum of a quadratic function by using "conjugate directions". We define the directions $\mathbf{p}$ and $\mathbf{q}$ in the space of the variables to be conjugate with respect to the negative definite (the condition of negative definiteness is needed to provide a solution to the optimization problem) quadratic objective function:

$$(15) \qquad \Phi(x_1, x_2, \cdots, x_n) = c + \sum_{i=1}^{n} a_i x_i + \tfrac{1}{2} \sum_{i=1}^{n} \sum_{j=1}^{n} G_{ij} x_i x_j$$

if they are both nonzero, and if they satisfy the equation

$$(16) \qquad \sum_{i=1}^{n} \sum_{j=1}^{n} p_i G_{ij} q_j = 0.$$

The reason they are useful is that if we search in the direction $\mathbf{p}$, and find the point $(x_1^{(k)}, x_2^{(k)}, \cdots, x_n^{(k)})$ that maximizes $\Phi(x_1^{(k)}, x_2^{(k)}, \cdots, x_n^{(k)})$, and then we search from $(x_1^{(k)}, x_2^{(k)}, \cdots, x_n^{(k)})$ in the conjugate direction $\mathbf{q}$ to reach the new estimate $(x_1^{(k+1)}, x_2^{(k+1)}, \cdots, x_n^{(k+1)})$, then the new value of the objective function cannot be increased by immediately searching again in the direction $\mathbf{p}$. (This statement is proved by straightforward algebra based on the observation that $(x_1, x_2, \cdots, x_n)$ is a best point in the direction $\mathbf{p}$ if and only if the gradient of the objective function at $(x_1, x_2, \cdots, x_n)$ is orthogonal to $\mathbf{p}$.) Using this remark, if we are given n mutually conjugate directions $\mathbf{d}_1, \mathbf{d}_2, \cdots, \mathbf{d}_n$, then we can calculate the exact maximum of the quadratic function $\Phi(x_1, x_2, \cdots, x_n)$ by the following process.

We start with an arbitrary estimate $(x_1^{(1)}, x_2^{(1)}, \cdots, x_n^{(1)})$ of the required variables, and then we search along the n conjugate directions. Specifically, for $k = 1, 2, \cdots, n$, the kth search replaces the estimate $(x_1^{(k)}, x_2^{(k)}, \cdots, x_n^{(k)})$ by the estimate:

(17) $$(x_1^{(k+1)}, x_2^{(k+1)}, \cdots, x_n^{(k+1)}) = (x_1^{(k)}, x_2^{(k)}, \cdots, x_n^{(k)}) + \lambda_k \mathbf{d}_k,$$

where λ_k is calculated to maximize the value of $\Phi(x_1^{(k+1)}, x_2^{(k+1)}, \cdots, x_n^{(k+1)})$. At the final point $(x_1^{(n+1)}, x_2^{(n+1)}, \cdots, x_n^{(n+1)})$ the gradient of the objective function must be orthogonal to the n directions $\mathbf{d}_1, \mathbf{d}_2, \cdots; \mathbf{d}_n$, but (16) can be used to show that mutually conjugate directions are independent. Therefore the n linear searches find the position of the optimum.

To use these ideas to optimize a given quadratic function is straightforward if the second derivative matrix is given, because one just has to calculate a set of n mutually conjugate directions to satisfy conditions like (16). In this case it is usual to choose $\mathbf{d}_1$ first, and then $\mathbf{d}_2$, and so on, in which case, if we suppose that the length of a direction is unimportant, there are $n - k$ degrees of freedom in the choice of $\mathbf{d}_k$. These degrees of freedom give many different methods, which are related to the various conjugate gradient algorithms for solving linear systems [22]. However, usually the objective function is not quadratic, so the concept of conjugacy is not defined. Therefore we must look for ways of simulating the condition (16) that will yield exact conjugacy in the quadratic case.

Zoutendijk [40] suggests one method that has not been used much, but it could prove to be a very valuable idea. He remarks that if the search direction $\mathbf{p}$ is used to move from the estimate $(x_1^{(k)}, x_2^{(k)}, \cdots, x_n^{(k)})$ to the estimate $(x_1^{(k+1)}, x_2^{(k+1)}, \cdots, x_n^{(k+1)})$, then, in the case that the objective function is quadratic, the condition (16) is the same as the equation

(18) $$\sum_{i=1}^{n} (g_i^{(k+1)} - g_i^{(k)})q_i = 0,$$

where $g_i^{(k)}$ is the ith component of the gradient of the objective function at $(x_1^{(k)}, x_2^{(k)}, \cdots, x_n^{(k)})$. Equation (18) is valuable because it contains no explicit second derivative terms, so we have a means of obtaining conjugacy when only values and first derivatives of $F(x_1, x_2, \cdots, x_n)$ are available. For instance we could start with an arbitrary search direction $\mathbf{d}_1$, and then, for $k = 2, 3, \cdots, n$, we could calculate the search direction $\mathbf{d}_k$ to be orthogonal to the changes in the gradient vector that were caused by the moves in the directions $\mathbf{d}_1, \mathbf{d}_2, \cdots, \mathbf{d}_{k-1}$.

The method just described simplifies in a remarkable and useful way if the first search direction is the direction of steepest ascent, and the later search directions $\mathbf{d}_k$, $k = 2, 3, \cdots$, are chosen to be the gradient $(g_1^{(k)}, g_2^{(k)}, \cdots, g_n^{(k)})$ plus the appropriate linear combination of the previous directions $\mathbf{d}_1, \mathbf{d}_2, \cdots, \mathbf{d}_{k-1}$, that provides conjugacy when $F(x_1, x_2, \cdots, x_n)$ is quadratic. In this case the required mutual conjugacy of the search directions can be obtained simply by choosing $\mathbf{d}_k$ to be of the form:

(19) $$\mathbf{d}_k = (g_1^{(k)}, g_2^{(k)}, \cdots, g_n^{(k)}) + \beta_{k-1} \mathbf{d}_{k-1},$$

where β_{k-1} is a real number. Moreover, β_{k-1} can be calculated from the formula

(20) $$\beta_{k-1} = \sum_{i=1}^{n} [g_i^{(k)}]^2 \bigg/ \sum_{i=1}^{n} [g_i^{(k-1)}]^2,$$

because it happens that, in the quadratic case, the process causes the successive gradients $(g_1^{(k)}, g_2^{(k)}, \cdots, g_n^{(k)})$, $k = 1, 2, \cdots$, to be mutually orthogonal. This method is the Fletcher–Reeves [15] algorithm, and it has been used successfully on many problems. The iteration is so elegant that it is straightforward to apply it any number of times when the objective function is not quadratic. However, its success on a quadratic function depends on the first iteration being a steepest ascent step, and after n iterations the identity of the first step must be lost. Therefore, to obtain a fast rate of convergence, Fletcher and Reeves recommend that the algorithm should be restarted with a steepest ascent step after every $n + 1$ iterations. An important advantage of the method, which is not obtained by other conjugate direction algorithms, is that it does not require storage space for any $n \times n$ matrices.

Conjugate directions have also been used to construct a successful algorithm [29] that does not require the explicit evaluation of any derivatives. When derivatives are not available conjugate directions are generated by using the following fact: if two separate points, say, $(y_1, y_2, \cdots, y_n)$ and $(z_1, z_2, \cdots, z_n)$, in the space of the variables are both obtained by searching along a direction, say $\mathbf{p}$, for the maximum value of the objective function, and if the objective function is quadratic, then the direction

$$(21) \qquad \mathbf{q} = (y_1, y_2, \cdots, y_n) - (z_1, z_2, \cdots, z_n)$$

is conjugate to $\mathbf{p}$. (This property is also used in [28], [32] and [33].) An iteration of Powell's algorithm requires an estimate $(x_1^{(k)}, x_2^{(k)}, \cdots, x_n^{(k)})$ of the position of the optimum, and n independent search directions, say $\mathbf{d}_1^{(k)}, \mathbf{d}_2^{(k)}, \cdots, \mathbf{d}_n^{(k)}$, which initially are chosen to be the coordinate directions. Each iteration can change both the estimate and the search directions, and the algorithm is designed so that, if it is applied to a negative definite quadratic function, then usually n iterations will find the optimum, because the search directions become mutually conjugate. An important feature of an iteration is that special measures are taken to prevent the n search directions from becoming linearly dependent, but we omit these measures from our description of an iteration in order to make the main idea more easy to understand.

The simplified iteration begins by searching along each of the directions in turn, and thus the estimate $(x_1^{(k)}, x_2^{(k)}, \cdots, x_n^{(k)})$ is changed by the amount

$$(22) \qquad \boldsymbol{\delta} = \sum_{i=1}^{n} \lambda_i \mathbf{d}_i^{(k)},$$

where, for $i = 1, 2, \cdots, n$, the multiplier λ_i is calculated to maximize the intermediate value of the objective function

$$(23) \qquad F\left\{(x_1^{(k)}, x_2^{(k)}, \cdots, x_n^{(k)}) + \sum_{j=1}^{i} \lambda_i \mathbf{d}_j^{(k)}\right\}.$$

Then the displacement $\boldsymbol{\delta}$, defined by (22), is also used as a search direction, and the resultant point, say $(x_1^{(k+1)}, x_2^{(k+1)}, \cdots, x_n^{(k+1)})$, is the starting estimate for the next iteration Finally the search directions are replaced by the set

(24)
$$d_i^{(k+1)} = d_{i+1}^{(k)}, \qquad\qquad i = 1, 2, \cdots, n-1,$$
$$d_n^{(k+1)} = \delta.$$

The result of applying this process to a quadratic function is that usually the new directions generated by the kth iteration, namely $d_1^{(k+1)}, d_2^{(k+1)}, \cdots, d_n^{(k+1)}$, have the property that the last k of them are mutually conjugate. This remark is proved by induction [29], unless linear dependence occurs [39], when the special measures of the complete algorithm become essential. Some numerical comparisons of this method with direct search techniques have been made by Box [5] and by Fletcher [13].

The last conjugate direction algorithm that we shall consider is probably the most successful of all. It is due to Davidon [9], but, because the original description has limited circulation, the paper by Fletcher and Powell [14] is recommended. Originally the algorithm was called a "variable metric" method, and we shall describe it from this point of view, because it is rather difficult to visualize a description that is based on conjugate directions.

To introduce the description, we note that the search direction $(\delta_1, \delta_2, \cdots, \delta_n)$ of the steepest ascent iteration at the point $(x_1, x_2, \cdots, x_n)$ is given by the formula

(25)
$$\delta_i = \sum_{j=1}^{n} I_{ij} g_j, \qquad\qquad i = 1, 2, \cdots, n,$$

where I is the unit matrix, and g_j is the jth component of the gradient of the objective function (see (2)). Also we note that, if in place of the unit matrix in (25) we substitute any positive definite matrix, then we obtain another search direction that has the property that its angle with the gradient vector is less than ninety degrees, so, provided that the gradient is nonzero, a search along this new direction will increase the objective function. In particular, in view of the remarks made at the end of § 4, some choice of this positive definite matrix will provide the fast convergence of the Newton-Raphson iteration. Therefore the kth iteration of Davidon's method changes the estimate $(x_1^{(k)}, x_2^{(k)}, \cdots, x_n^{(k)})$ to the estimate $(x_1^{(k+1)}, x_2^{(k+1)}, \cdots, x_n^{(k+1)})$ by searching for the maximum of the objective function along the direction having the components:

(26)
$$\delta_i^{(k)} = \sum_{j=1}^{n} H_{ij}^{(k)} g_j^{(k)}, \qquad\qquad i = 1, 2, \cdots, n,$$

where $H^{(k)}$ is a positive definite matrix which is chosen with the intention of providing a good rate of convergence. The method uses only values and first derivatives of the objective function, and it accounts for second derivative terms by changing $H^{(k)}$ according to differences between the calculated gradients of $F(x_1, x_2, \cdots, x_n)$. Usually the initial matrix $H^{(1)}$ is chosen to be the unit matrix, but any other positive definite matrix may be used instead, except that its elements should not conflict severely with the scaling of the variables $x_1, x_2, \cdots, x_n$, because of the effect of computer roundoff errors [3].

To calculate $H^{(k+1)}$, the Davidon iteration adds a correction term to the matrix $H^{(k)}$, that depends on the two vectors:

(27)
$$\sigma^{(k)} = (x_1^{(k+1)}, x_2^{(k+1)}, \cdots, x_n^{(k+1)}) - (x_1^{(k)}, x_2^{(k)}, \cdots, x_n^{(k)}),$$
$$\gamma^{(k)} = (g_1^{(k+1)}, g_2^{(k+1)}, \cdots, g_n^{(k+1)}) - (g_1^{(k)}, g_2^{(k)}, \cdots, g_n^{(k)}).$$

This correction term is intended to make $H^{(k+1)}$ close to the matrix $-G^{-1}$, in order that the direction (26) shall be near to the Newton-Raphson direction (7). The use of the vectors (27) is guided by the fact that, if the objective function is quadratic, then the equation

(28)
$$\sigma^{(k)} = G^{-1}\gamma^{(k)}$$

holds, for $H^{(k+1)}$ is calculated to satisfy the equation

(29)
$$\sigma^{(k)} = -H^{(k+1)}\gamma^{(k)}.$$

Specifically the formula

(30)
$$H^{(k+1)} = H^{(k)} - \frac{\sigma^{(k)}\sigma^{(k)T}}{\sigma^{(k)T}\gamma^{(k)}} - \frac{H^{(k)}\gamma^{(k)}\gamma^{(k)T}H^{(k)}}{\gamma^{(k)T}H^{(k)}\gamma^{(k)}}$$

is used, where the superscript T indicates a row vector. Note that nearly all the other descriptions of Davidon's method treat the minimization, rather than the maximization, problem, in which case the signs of the first part of the correction term, and of the right-hand side of (29), are different.

This iteration has two important properties [14]. The first is that if $H^{(k)}$ is positive definite, then the matrix $H^{(k+1)}$ is also positive definite, and the second is that when the algorithm is applied to a quadratic function the directions of search are mutually conjugate. Therefore the method usually converges quickly. However, very occasionally the iterations make slow progress when the vectors $(x_1^{(k)}, x_2^{(k)}, \cdots, x_n^{(k)})$ are far from the solution, and sometimes this happens when a steepest ascent step would cause a substantial increase in the value of the objective function. Although in these cases the algorithm often manages to overcome the difficulty automatically, it is worrying that the inefficiency occurs. This remark typifies the fact that we do not really understand the algorithm, except in the trivial case when the objective function is quadratic.

Stewart [36] has extended Davidon's method, so that derivatives of the objective function are not needed. Instead the first derivatives are approximated by differences between function values, the main feature of Stewart's algorithm being that the intervals used in the difference approximations are chosen automatically, with a view to balancing truncation and roundoff errors. Numerical examples show that the resultant procedure is a little more efficient than Powell's [29] method when the number of variables is less than five, and for larger values of n the improvement is considerable.

8. Rank-one methods. This section describes and discusses just one recent idea that is likely to yield faster and more reliable algorithms for solving the unconstrained optimization problem, when values and first derivatives of the objective function are available. It has been stated by Broyden [7] and by Davidon [10], but they missed a property that makes the technique especially promising. This property was noticed by Wolfe [38].

The idea is a development of Davidon's variable metric algorithm, for it is derived by considering definitions of $H^{(k+1)}$ different from (30), that also satisfy (29). In particular there is just one way of calculating $H^{(k+1)}$ so that the difference $\{H^{(k+1)} - H^{(k)}\}$ is a symmetric matrix of rank one. It is the formula:

$$(31) \qquad H^{(k+1)} = H^{(k)} - \frac{\{\sigma^{(k)} + H^{(k)}\gamma^{(k)}\}\{\sigma^{(k)} + H^{(k)}\gamma^{(k)}\}^T}{\{\sigma^{(k)} + H^{(k)}\gamma^{(k)}\}^T\gamma^{(k)}},$$

and the idea is to calculate $H^{(k+1)}$ from $H^{(k)}$ by (31), where $\sigma^{(k)}$ and $\gamma^{(k)}$ are defined by (27).

To use (31) in an algorithm it is necessary to fix rules for calculating $(x_1^{(k+1)}, x_2^{(k+1)}, \cdots, x_n^{(k+1)})$ from $(x_1^{(k)}, x_2^{(k)}, \cdots, x_n^{(k)})$, and no doubt these rules will use the matrix $H^{(k)}$. The reason the idea is so valuable is that there are very many choices of rules such that n applications of (31) cause H to equal $-G^{-1}$ when the objective function is quadratic, and, because of the form of the Newton-Raphson iteration (7), this property can lead to fast convergence in the general case.

Both Broyden [7] and Davidon [10] follow (26) to define the rules for calculating $(x_1^{(k+1)}, x_2^{(k+1)}, \cdots, x_n^{(k+1)})$. They use the formula

$$(32) \qquad x_i^{(k+1)} = x_i^{(k)} + \alpha^{(k)} \sum_{j=1}^{n} H_{ij}^{(k)} g_j^{(k)},$$

where $\alpha^{(k)}$ is a parameter that is independent of i. Broyden proves that, if the choice of $\alpha^{(k)}$ is arbitrary, except that it must not cause $H^{(k+1)}$ to be singular or ill-defined (through the denominator of (31) being equal to zero), and if $F(x_1, x_2, \cdots, x_n)$ is quadratic, then $H^{(n+1)}$ will equal $-G^{-1}$. The important feature of this theorem is that it does not depend on calculating $\alpha^{(k)}$ by applying a one-dimensional search to maximize the objective function.

Davidon's new algorithm always uses $\alpha^{(k)} = 1$, so it can happen that expression (31) is not defined. However, if the vectors $\sigma^{(k)}$ and $\gamma^{(k)}$ are such that $H^{(k+1)}$ would not be positive definite, then Davidon defines $H^{(k+1)}$ by adding a different multiple of $\{\sigma^{(k)} + H^{(k)}\gamma^{(k)}\}\{\sigma^{(k)} + H^{(k)}\gamma^{(k)}\}^T$ to $H^{(k)}$, so that positive definiteness is obtained. Another feature of the method is that if $F(x_1^{(k+1)}, x_2^{(k+1)}, \cdots, x_n^{(k+1)})$ is less than $F(x_1^{(k)}, x_2^{(k)}, \cdots, x_n^{(k)})$, then, after $H^{(k+1)}$ has been calculated, $(x_1^{(k+1)}, x_2^{(k+1)}, \cdots, x_n^{(k+1)})$ is replaced by $(x_1^{(k)}, x_2^{(k)}, \cdots, x_n^{(k)})$ for the purposes of the next iteration.

The fact that linear searches are avoided is very valuable, because conjugate direction methods use many evaluations of the objective function just to locate optima along lines sufficiently accurately. Moreover, Box [5] shows by example that objective functions with well-defined solutions can be such that there are no optimal points along certain lines in the space of the variables, so the presence of linear searches can itself cause failure.

Wolfe's important observation [38] is that (31) can cause $H^{(n+1)}$ to equal $-G^{-1}$ in the quadratic case, without the restriction that $x_i^{(k+1)}$, $i = 1, 2, \cdots, n$, shall be calculated by (32). This statement is justified by the following argument.

If $F(x_1, x_2, \cdots, x_n)$ is the negative definite quadratic function (15), and if $\boldsymbol{\sigma}^{(k)}$ and $\boldsymbol{\gamma}^{(k)}$ are related by (27), then, for all vectors $\mathbf{u}$, the equation

$$(33) \qquad \{\boldsymbol{\sigma}^{(k)} + H^{(k)}\boldsymbol{\gamma}^{(k)}\}^T \mathbf{u} = \boldsymbol{\gamma}^{(k)T}\{G^{-1} + H^{(k)}\}\mathbf{u}$$

holds. Therefore if $\boldsymbol{\sigma}^{(k)}$ and $\boldsymbol{\gamma}^{(k)}$ are such that (31) does not have a zero denominator, and if $\mathbf{u}$ satisfies the equation

$$(34) \qquad G^{-1}\mathbf{u} = -H^{(k)}\mathbf{u},$$

then we deduce, from (31) and (33), that $\mathbf{u}$ also satisfies the equation

$$(35) \qquad G^{-1}\mathbf{u} = -H^{(k+1)}\mathbf{u}.$$

But for $j = 1, 2, \cdots, k$, the matrix $H^{(j+1)}$ is calculated to make the equation:

$$(36) \qquad G^{-1}\boldsymbol{\gamma}^{(j)} = -H^{(j+1)}\boldsymbol{\gamma}^{(j)}$$

hold, so from our remarks we obtain the result:

$$(37) \qquad G^{-1}\boldsymbol{\gamma}^{(j)} = -H^{(k+1)}\boldsymbol{\gamma}^{(j)}, \qquad\qquad j = 1, 2, \cdots, k.$$

It follows that if, for $k = 1, 2, \cdots, n$, the vectors $\boldsymbol{\sigma}^{(k)}$ are independent (which implies the independence of $\boldsymbol{\gamma}^{(k)}$, $k = 1, 2, \cdots, n$), and if these vectors are calculated so that (31) is well-defined, then $H^{(n+1)}$ is equal to $-G^{-1}$. Thus all the second derivative information is obtained without any linear searching, and we note that there is much freedom in the choice of $\boldsymbol{\sigma}^{(k)}$, $k = 1, 2, \cdots, n$.

Some ways of exploiting this idea are suggested in § 10.

Note that n applications of (31) is not the same as solving a large system of linear equations to determine the coefficients of a quadratic function, because solving the equations does not automatically cause the matrix $H^{(n+1)}$ to be symmetric.

9. Poor initial estimates. If the initial vector of variables $(x_1^{(1)}, x_2^{(1)}, \cdots, x_n^{(1)})$ is far from the position of the maximum of $F(x_1, x_2, \cdots, x_n)$, and if a conjugate direction algorithm is applied to solve the optimization algorithm, then it usually happens that most of the computer time is spent on obtaining a point that is close to the solution, rather than on completing the calculation after a good estimate has been obtained. Therefore the behavior of methods when $(x_1^{(k)}, x_2^{(k)}, \cdots, x_n^{(k)})$ is far from a maximum is of interest, but it has been studied very little. Some progress has been made by Greenstadt [21], and by Goldfeld, Quandt and Trotter [18], for they have suggested algorithms that, unlike the Newton-Raphson method, will act sensibly when the estimate $(x_1^{(k)}, x_2^{(k)}, \cdots, x_n^{(k)})$ is so poor that the matrix of second derivatives of the objective function is not negative definite. We consider the two methods in this section.

Greenstadt's method [21] is a simple extension of the Newton-Raphson method. The kth iteration calculates the second derivative matrix $G^{(k)}$ at the estimate $(x_1^{(k)}, x_2^{(k)}, \cdots, x_n^{(k)})$, and then an eigenvalue analysis obtains the reduction

$$(38) \qquad G^{(k)} = \Omega\Lambda\Omega^T,$$

where Ω is orthogonal and Λ is diagonal. In place of G^{-1} in (7), Greenstadt uses the matrix $(\Omega\Lambda^*\Omega^T)^{-1}$ to calculate the search direction $(\delta_1^{(k)}, \delta_2^{(k)}, \cdots, \delta_n^{(k)})$, where Λ^* is the diagonal matrix obtained by replacing the diagonal elements of Λ by the negatives of their moduli. The estimate $(x_1^{(k+1)}, x_2^{(k+1)}, \cdots, x_n^{(k+1)})$ is calculated by searching from $(x_1^{(k)}, x_2^{(k)}, \cdots, x_n^{(k)})$ in the direction $(\delta_1^{(k)}, \delta_2^{(k)}, \cdots, \delta_n^{(k)})$ for the maximum of the objective function. Although the original description is not supported by numerical examples, it is stated that the method works very well in practice. Conversations with various people have confirmed this assertion.

We have already met the main idea of the algorithm due to Goldfeld, Quandt and Trotter [18], for it is expressed by (12). The nonnegative parameter α is calculated so that the matrix $[\alpha I - G]$ is positive definite, the actual value of α being equal to the maximum of zero and R plus the largest eigenvalue of G, where R is a positive number that is adjusted automatically. One purpose of R is to control the length of the correction vector $(\delta_1, \delta_2, \cdots, \delta_n)$ (calculated by (12)), so that the value of the objective function at $(x_1^{(k)} + \delta_1^{(k)}, x_2^{(k)} + \delta_2^{(k)}, \cdots, x_n^{(k)} + \delta_n^{(k)})$ is larger than $F(x_1^{(k)}, x_2^{(k)}, \cdots, x_n^{(k)})$.

The need to calculate second derivatives and the need to solve the eigenvalue problem both detract from the utility of the above two algorithms. However, because of the lack of sensible strategies when the estimate $(x_1^{(k)}, x_2^{(k)}, \cdots, x_n^{(k)})$ is far from the solution, it can be worthwhile to study such methods. In particular, by experimenting with different algorithms that require second derivatives, efficient ideas may be identified, and only then may it be worthwhile to spend much ingenuity on avoiding the explicit calculation of G.

A good theoretical reason for their method is given by Goldfeld, Quandt and Trotter [18], and the same reason was noticed independently and earlier by Marquardt [26]. It applies when the objective function is quadratic, and when α exceeds the greatest eigenvalue of G. In this case the correction $(\delta_1, \delta_2, \cdots, \delta_n)$, defined by (12), has the property that $F(x_1 + \delta_1, x_2 + \delta_2, \cdots, x_n + \delta_n)$ is the greatest value of the objective function within the sphere of points whose Euclidean distance from $(x_1, x_2, \cdots, x_n)$ does not exceed $(\delta_1^2 + \delta_2^2 + \cdots + \delta_n^2)^{1/2}$. By choosing positive definite matrices that are different from I in (12), we can calculate best points within ellipsoids, rather than within spheres. These extensions are discussed in [19].

10. Conclusions. We have traced the main ideas of unconstrained optimization, with the intention of identifying successful algorithms, and of setting the scene for future research. Our main conclusions on the algorithms that have been in use for a number of years are that occasionally a direct search method is preferable, but usually it is better to apply Davidon's method [9] (if first derivatives are available), Powell's method [29] or Stewart's method [36] (if the user prefers not to program the calculation of derivatives). The number of variables that can be handled by these techniques depends strongly on $F(x_1, x_2, \cdots, x_n)$, a rough guide being that Davidon's algorithm has been successful for $n = 100$, and the other two methods have solved problems with twenty variables. The numerical evidence suggests that Stewart's technique becomes much more efficient than Powell's as the number of variables increases. However, if $F(x_1, x_2, \cdots, x_n)$ is not differentiable the suitability of conjugate direction methods is doubtful. In this case it is often necessary to study the properties of the objective function, in order to identify special features that will help locate the solution; but to avoid unnecessary effort it is prudent to try an automatic search method first, just in case it is successful.

Our conclusion on the theory of the algorithms is that a great deal of work needs to be done. For instance, in the cases when Davidon's method makes slow progress (although the gradient of $F(x_1, x_2, \cdots, x_n)$ is substantial), and then manages to take large steps again, it is not known whether the improvement is due to computer rounding errors, or whether it would occur if exact arithmetic were used. Also conditions for convergence and rates of convergence for nearly all the algorithms described in §§ 6–9 have not been found. It is hoped that this lack of theory will stimulate some research.

The most exciting prospect is the potential of the rank-one methods, described in § 8, because very many choices of displacements $(x_1^{(k+1)} - x_1^{(k)}, x_2^{(k+1)} - x_2^{(k)}, \cdots, x_n^{(k+1)} - x_n^{(k)})$ provide the exact solution when the objective function is quadratic. For instance, this freedom can be used to ensure that the successive displacements do not tend to lie in just a part of the full space of the variables, which is necessary to cause $H^{(k)}$ to converge to G^{-1}. Also this freedom permits Levenberg's idea [25], given by (12), to be exploited. Since these benefits can probably be obtained without the need for linear searches, it is likely that the current methods of unconstrained optimization will be replaced by much better algorithms.

REFERENCES

[1] HIROTUGU AKAIKE, *On a successive transformation of probability distribution and its application to the analysis of the optimum gradient method*, Ann. Inst. Statist. Math. Tokyo, 11 (1959), pp. 1–16.

[2] ROBERT M. BAER, *Note on an extremum locating algorithm*, Comput. J., 5 (1962), p. 193.

[3] YONATHON BARD, *On a numerical instability of Davidon-like methods*, IBM New York Scientific Center Rep. no. 320–2913, 1967.

[4] A. D. BOOTH, *Numerical Methods*, Butterworths, London, 1957.

[5] M. J. BOX, *A comparison of several current optimization methods, and the use of transformations in constrained problems*, Comput. J., 9 (1966), pp. 67–77.

[6] M. J. BOX, D. DAVIES AND W. H. SWANN, *Optimization Techniques*, I.C.I. Monograph no. 5, Oliver and Boyd Ltd., Edinburgh and London, 1968, to appear.

[7] C. G. BROYDEN, *Quasi-Newton methods and their application to function minimization*, Math. Comp., 21 (1967), pp. 368–381.

[8] A. CAUCHY, *Methode générale pour la résolution des systemes d'équations simultanées*, C.R. Acad. Sci. Paris, 25 (1847), p. 536.

[9] W. C. DAVIDON, *Variable metric method for minimization*, A.E.C. Research and Development Rep. ANL-5990 (rev.), 1959.

[10] ———, *Variance algorithm for minimization*, Comput. J., 10 (1968), pp. 406–410.

[11] A. W. DICKINSON, *Nonlinear optimization: some procedures and examples*, Proc. 19th ACM National Conference, 1964, pp. E1.2/1–E1.2/8.

[12] D. K. FADEEV AND V. N. FADEEVA, *Computational Methods of Linear Algebra*, W. H. Freeman, San Francisco, 1963.

[13] R. FLETCHER, *Function minimization without evaluating derivatives—a review*, Comput. J., 8 (1965), pp. 33–41.

[14] R. FLETCHER AND M. J. D. POWELL, *A rapidly convergent descent method for minimization*, Ibid., 6 (1963), pp. 163–168.

[15] R. FLETCHER AND C. M. REEVES, *Function minimization by conjugate gradients*, Ibid., 7 (1964), pp. 149–154.

[16] G. E. FORSYTHE, *On the asymptotic directions of the s-dimensional optimum gradient method*, Numer. Math., 11 (1968), pp. 57–76.

[17] G. E. FORSYTHE AND T. S. MOTZKIN, *Asymptotic properties of the optimum gradient method*, Bull. Amer. Math. Soc., 57 (1951), p. 183.

[18] S. M. GOLDFELD, R. E. QUANDT AND H. F. TROTTER, *Maximization by quadratic hill-climbing*, Econometrica, 34 (1966), pp. 541–551.

[19] ———, *Maximization by improved quadratic hill-climbing and other methods*, Econometric Research Memo. 95, Princeton University, 1968.

[20] A. A. GOLDSTEIN, *Cauchy's method of minimization*, Numer. Math., 4 (1962), pp. 146–150.

[21] JOHN GREENSTADT, *On the relative efficiencies of gradient methods*, Math. Comp., 21 (1967), pp. 360–367.

[22] M. R. HESTENES AND E. STIEFEL, *Methods of conjugate gradients for solving linear systems*, J. Res., Nat. Bur. Standards, 49 (1952), pp. 409–436.

[23] R. HOOKE AND T. A. JEEVES, *Direct search solution of numerical and statistical problems*, J. Assoc. Comput. Mach., 8 (1961), pp. 212–221.

[24] ALBERTO LEON, *A comparison among eight known optimizing procedures*, Recent Advances in Optimization Techniques, A. Lavi and T. P. Vogl, eds., John Wiley, New York, 1966.

[25] K. LEVENBERG, *A method for the solution of certain non-linear problems in least squares*, Quart. Appl. Math., 2 (1944), pp. 164–168.

[26] DONALD W. MARQUARDT, *An algorithm for least-squares estimation of nonlinear parameters*, J. Soc. Indust. Appl. Math., 11 (1963), pp. 431–441.

[27] J. A. NELDER AND R. MEAD, *A simplex method for function minimization*, Comput. J., 7 (1965), pp. 308–313.

[28] M. J. D. POWELL, *An iterative method for finding stationary values of functions of several variables*, Ibid., 5 (1962), pp. 147–151.

[29] ———, *An efficient method for finding the minimum of a function of several variables without calculating derivatives*, Ibid., 7 (1964), pp. 155–162.

[30] ———, *Minimization of functions of several variables*, Numerical Analysis: An Introduction, J. Walsh, ed., Academic Press, London, 1966, pp. 143–157.

[31] H. H. ROSENBROCK, *An automatic method for finding the greatest or the least value of a function*, Comput. J., 3 (1960), pp. 175–184.

[32] B. V. SHAH, R. J. BEUHLER AND O. KEMPTHORNE, *Some algorithms for minimizing a function of several variables*, J. Soc. Indust. Appl. Math., 12 (1964), pp. 74–92.

[33] C. S. SMITH, *The automatic computation of maximum likelihood estimates*, Rep. S.C. 846/MR/40, N.C.B. Scientific Dept., 1962.

[34] H. A. SPANG, III, *A review of minimization techniques for nonlinear functions*, this Review, 4(1962), pp. 343–365.

[35] W. SPENDLEY, G. R. HEXT AND F. R. HIMSWORTH, *Sequential application of simplex designs in optimization and evolutionary operation*, Technometrics, 4 (1962), p. 441.

[36] G. W. STEWART, III, *A modification of Davidon's minimization method to accept difference approximations of derivatives*, J. Assoc. Comput. Mach., 14 (1967), pp. 72–83.

[37] W. H. SWANN, *Report on the development of a new direct search method of optimization*, Research Note 64/3, Central Instrument Laboratory, I.C.I. Ltd., 1964.

[38] P. WOLFE, *Another variable metric method*, Working paper, 1967.

[39] WILLARD I. ZANGWILL, *Minimizing a function without calculating derivatives*, Comput. J., 10 (1967), pp. 293–296.

[40] G. ZOUTENDIJK, *Methods of Feasible Directions*, Elsevier, Amsterdam, 1960.

II. MATHEMATICAL PROGRAMMING

2. ELEMENTS OF LARGE-SCALE MATHEMATICAL PROGRAMMING

PART I: CONCEPTS*†‡§

ARTHUR M. GEOFFRION

University of California, Los Angeles

A framework of concepts is developed which helps to unify a substantial portion of the literature on large-scale mathematical programming. These concepts fall into two categories. The first category consists of problem manipulations that can be used to derive what are often referred to as "master" problems; the principal manipulations discussed are Projection, Inner Linearization, and Outer Linearization. The second category consists of solution strategies that can be used to solve the master problems, often with the result that "subproblems" arise which can then be solved by specialized algorithms. The Piecewise, Restriction, and Relaxation strategies are the principal ones discussed. Numerous algorithms found in the literature are classified according to the manipulation/strategy pattern they can be viewed as using, and the usefulness of the framework is demonstrated by using it (see Part II of this paper) to rederive a representative selection of algorithms.

The material presented is listed in the following order: The first section is introductory in nature, and discusses types of large-scale problems, the scope of discussion and the literature, and the notation used. The second section, entitled "Problem Manipulations: Source of 'Master' Problems" covers the subjects of projection, inner linearization and outer linearization. The third section, "Solution Strategies: Source of 'Subproblems'," discusses piecewise strategy, restriction and relaxation. The fourth section is entitled "Synthesizing Known Algorithms from Manipulations and Strategies," and is followed by a concluding section and an extensive bibliography.

1. Introduction

The development of efficient optimization techniques for large structured mathematical programs is of major significance in economic planning, engineering, and management science. A glance at the bibliography of this paper will reveal the magnitude of the effort devoted to the subject in recent years. The purpose of this paper is to suggest a unifying framework to help both the specialist and nonspecialist cope with this rapidly growing body of knowledge.

The proposed framework is based on a relative handful of fundamental concepts that can be classified into two groups: *problem manipulations* and *solution strategies*. Problem manipulations are devices for restating a given problem in an alternative form that

* Received April 1969.

† This is the ninth in a series of twelve expository papers commissioned jointly by the Office of Naval Research and the Army Research Office under contract numbers Nonr-4004(00) and DA 49-092-ARO-16, respectively.

‡ Although this paper was originally prepared for publication in a single installment, an editorial decision was made to divide it into two parts because of its exceptional length. The reader may simply ignore this division, which has not occasioned any significant changes in the body of the text. All references are found at the end of Part II, which appears next in this issue.

§ Support for this work was provided by the Ford Foundation under a Faculty Research Fellowship, by the National Science Foundation under Grant GP-8740, and by the United States Air Force under Project RAND. It is a pleasure to acknowledge the helpful comments of A. Feinberg, B. L. Fox, and C. A. Holloway.

is apt to be more amenable to solution. The result is often what is referred to in the literature as a "master" problem. Dualization of a linear program is one familiar example of such a device. §2 discusses three others: Projection, Inner Linearization, and Outer Linearization. Solution strategies, on the other hand, reduce an optimization problem to a related sequence of simpler optimization problems. This often leads to "subproblems" amenable to solution by specialized methods. The Feasible Directions strategy is a well-known example, and §3 discusses the Piecewise, Restriction, and Relaxation strategies. The reader is probably already familiar with special cases of most of these concepts, if not with the names used for them here; the new terminology is introduced to emphasize the generality of the ideas involved.

By assembling these and a few other problem manipulations and solution strategies in various patterns, one can rederive the essential aspects of most known large-scale programming algorithms and even design new ones. §4 (see Part II of this paper) illustrates this for Benders Decomposition, Dantzig-Wolfe Decomposition, Rosen's Primal Partition Programming method, Takahashi's "local" approach, and a procedure recently devised by the author for nonlinear decomposition.

Although much of the presentation is elementary, for full appreciation the reader will find it necessary to have a working knowledge of the theory and computational methods of linear and nonlinear programming about at the level of a first graduate course in each subject.

1.1 *Types of Large-Scale Problems*

It is important to realize that size alone is not the distinguishing attribute of the field of "large-scale programming," but rather size in conjunction with *structure*. Large-scale programs almost always have distinctive and pervasive structure beyond the usual convexity or linearity properties. The principal focus of large-scale programming is the exploitation of various special structures for theoretical and computational purposes.

There are, of course, many possible types of structure. Among the commonest and most important general types are these: multidivisional, combinatorial, dynamic, and stochastic. *Multidivisional* problems consist of a collection of interrelated "subsystems" to be optimized.[1] The subsystems can be, for example, modules of an engineering system, reservoirs in a water resources system, departments or divisions of an organization, production units of an industry, or sectors of an economy. *Combinatorial* problems typically have a large number of variables because of the numerous possibilities for selecting routes, machine setups, schedules, etc.[2] Problems with *dynamic* aspects grow large because of the need to replicate constraints and variables to account for a number of time periods.[3] And problems with *stochastic* or uncertainty aspects are often larger than they would otherwise be in order to account for alternative possible realizations of imperfectly known entities.[4] A method that successfully exploits one specific

---

[1] See, e.g., Aoki 68, Bradley 67, Gould 59, Hass 68, Kornai and Liptak 65, Lasdon and Schoeffler 66, Malinvaud 67, Manne and Markowitz 63, Parikh and Shephard 67, Rosen and Ornea 63, Tcheng 66.

[2] See, e.g., Appelgren 69, Dantzig 60, Dantzig, Blattner and Rao 67, Dantzig, Fulkerson and Johnson 54, Dantzig and Johnson 64, Ford and Fulkerson 58, Gilmore and Gomory 61, 63, and 65, Glassey 66, Held and Karp 69, Midler and Wollmer 69, Rao and Zionts 68.

[3] See, e.g., Charnes and Cooper 55, Dantzig 55b, 59, Dzielinski and Gomory 65, Glassey 68, Rao 68, Robert 63, Rosen 67, Van Slyke and Wets 69, Wagner 57, Wilson 66.

[4] See, e.g., Dantzig and Madansky 61, El Agizy 67, Van Slyke and Wets 69, Wolfe and Dantzig 62.

structure can usually be adapted to exploit other specific structures of the same general type. Perhaps needless to say, problems are not infrequently encountered which fall simultaneously into two or more of these general categories.

The presence of a large number of variables or constraints can be due not only to the intrinsic nature of a problem as suggested above, but also to the chosen representation of the problem. Sometimes a problem with a few nonlinearities, for example, is expressed as a completely linear program by means of piecewise-linear or tangential linear approximation to the nonlinear functions or sets (cf. §§2.2, 2.3). Such approximations usually greatly enlarge the size of the problem.[5]

1.2 *Scope of Discussion and the Literature*

The literature on the computational aspects of large-scale mathematical programming can be roughly dichotomized as follows:

I. Work aimed at improving the computational efficiency of a known solution technique (typically the Simplex Method) for special types of problems.

II. Work aimed at developing fundamentally new solution techniques.

The highly specialized nature of the category I literature and the availability of several excellent surveys thereon leave little choice but to focus this paper primarily on category II. Fortunately this emphasis would be appropriate anyway, since category II is far more amorphous and in need of clarification.

Category I. The predominant context for category I contributions is the Simplex Method for linear programming. The objective is to find, for various special classes of problems, ways of performing each Simplex iteration in less time or using less primary storage. This work is in the tradition of the early and successful specialization of the Simplex Method for transportation problems and problems with upper-bounded variables. The two main approaches may be called *inverse compactification* and *mechanized pricing*.

Inverse compactification schemes involve maintaining the basis inverse matrix or an operationally sufficient substitute in a more advantageous form than the explicit one. One of the earliest and most significant examples is the "product form" of the inverse [Dantzig and Orchard-Hays 54], which takes advantage of the sparseness of most large matrices arising in application. Other schemes involve triangular factorization, partitioning, or use of a "working basis" that is more tractable than the true one. See part A of Table 1. A survey of many such contributions is found in §II of [Dantzig 68]. The interested reader should also consult [Willoughby 69] which, in the course of collecting a number of recent advances in the methods of dealing with sparse matrices, points out much pertinent work done in special application areas such as engineering structures, electrical networks, and electric power systems. Well over a hundred references are given.

Mechanized pricing, sometimes called *column generation*, involves the use of a subsidiary optimization algorithm instead of direct enumeration to find the best nonbasic variable to enter the basis when there are many variables.[6] The first contribution of this sort was [Ford and Fulkerson 58], in which columns were generated by a network

---

[5] See, e.g., Charnes and Lemke 54, Gomory and Hu 62, Kelley 60.

[6] It is also possible to mechanize the search for the exiting basic variable when there are many constraints (e.g., Gomory and Hu 62, §4) or when what amounts to the Dual Method is used (e.g., §3 of Gomory and Hu 62, Abadie and Williams 63, Whinston 64, and part A of Table 2).

flow algorithm. Subsequent authors have proposed generating columns by other network algorithms, dynamic programming, integer programming, and even by linear programming itself. See part B of Table 1. Excellent surveys of such contributions are [Balinski 64] and [Gomory 63].

Category I contributions of comparable sophistication are relatively rare in the literature on nonlinear problems. It has long been recognized that it is essential to take advantage of the recursive nature of most of the computations; that is, one should obtain the data required at each iteration by economically updating the data available from the previous iteration, rather than by operating each time on the original problem data. In Rosen's gradient projection algorithm, for example, the required projection matrix is updated at each iteration rather than computed ab initio. This is quite different, however, from "compacting" the projection matrix for a particular problem structure, or "mechanizing" the search for the most negative multiplier by means of a subsidiary optimization algorithm. Little has been published along these lines (see, however, p. 153ff. and §8.3 of [Fiacco and McCormick 68] and [Rutenberg 70]). Of course, many nonlinear algorithms involve a sequence of derived linear programs and therefore can benefit from the techniques of large-scale linear programming.

Category II. We turn now to work aimed at developing new solution techniques for various problem structures—the portion of the literature to which our framework of fundamental concepts is primarily addressed.

As mentioned above, the fundamental concepts are of two kinds: problem manipulations and solution strategies. The key problem manipulations (§2) are Dualization, Projection, Inner Linearization, and Outer Linearization, while the key solution strategies (§3) are Feasible Directions, Piecewise, Restriction and Relaxation. These building block concepts can be used to reconstruct many of the existing computational proposals. Using Projection followed by Outer Linearization and Relaxation, for example, we can obtain Benders' Partitioning Procedure. Rosen's Primal Partition Programming algorithm can be obtained by applying Projection and then the Piecewise strategy. Dantzig-Wolfe Decomposition employs Inner Linearization and Restriction. Similarly, many other existing computational proposals for large-scale programming can be formulated as particular patterns of problem manipulations and solution strategies applied to a particular structure.

See Table 2 for a classification of much of the literature of category II in terms of such patterns. One key or representative paper from each pattern is italicized to

TABLE 1

Some Work Aimed at Improving the Efficiency of the
Simplex Method for Large-Scale Problems

A. *Inverse Compactification*
   Dantzig and Orchard-Hays 54; Dantzig 55a, 55b, 63b; Markowitz 57; Dantzig, Harvey, and McKnight 64; Heesterman and Sandee 65; Kaul 65; Bakes 66; Bennett 66; Bennett and Green 66; Saigal 66; Dantzig and Van Slyke 67; Sakarovitch and Saigal 67; Grigoriadis 69; Willoughby 69.
B. *Mechanized Pricing*†
   Ford and Fulkerson 58; Dantzig 60; Gilmore and Gomory 61,‡ 63, 65; Dantzig and Johnson 64; Bradley 65, Sec. 3; Glassey 66; Tomlin 66; Dantzig, Blattner and Rao 67; Elmaghraby 68; Lasdon and Mackey 68; Rao 68, Sec. II; Rao and Zionts 68; Graves, Hatfield and Whinston 69; Fox 69; Held and Karp 69, Sec. 4.

   † Most of the references in part C of Table 2 also use mechanized pricing.
   ‡ Discussed in Sec. 3.2.

TABLE 2

Classification of Some References by Pattern:
Problem Manipulation(s)/Solution Strategy

A. *Projection, Outer Linearization/Relaxation*

Benders 62; Balinski and Wolfe 63; Gomory and Hu 64, pp. 351–354; Buzby, Stone and Taylor 65; Weitzman 67; Geoffrion 68b, Sec. 3; Van Slyke and Wets 69, Sec. 2; Geoffrion 70.

B. *Projection/Piecewise*

Rosen 63, 64; Rosen and Ornea 63; Beale 63; Gass 66; Varaiya 66; Chandy 68; Geoffrion 68b, Sec. 5; Grigoriadis and Walker 68.

C. *Inner Linearization/Restriction*

Dantzig and Wolfe 60; Dantzig and Madansky 61, p. 175; Williams 62; Wolfe and Dantzig 62; Dantzig 63a, Ch. 24; Baumol and Fabian 64; Bradley 65, Sec. 2; Dzielinski and Gomory 65; Madge 65; Tcheng 66; Tomlin 66; Whinston 66; Malinvaud 67, Sec. V; Parikh and Shephard 67; Elmaghraby 68; Hass 68; Rao 68, Sec. III; Appelgren 69; Robers and Ben-Israel 70.

D. *Projection/Feasible Directions*

Zschau 67; Abadie and Sakarovitch 67; Geoffrion 68b, Sec. 4; Silverman 68; Grinold 69, Secs. IV and V.

E. *Dualization/Feasible Directions*

Uzawa 58; Takahashi 64, "local" approach; Lasdon 64, 68; Falk 65, 67; Golshtein 66; Pearson 66; Wilson 66; Bradley 67 (Sec. 3.2), 68 (Sec. 4); Grinold 69, Sec. III.

---

signify that it is discussed in some detail in §4. Familiarity with one such paper from each pattern should enable the reader to assimilate the other papers, given an understanding of the fundamental concepts at the level of §§2 and 3.

Table 2 does not pretend to embrace the whole literature of category II. There undoubtedly are other papers that can naturally be viewed in terms of the five patterns of Table 2, and there certainly are papers employing other patterns.[7] Sections 2 and 3 mention other papers that can be viewed naturally in terms of one of the problem manipulations or solution strategies discussed there. Still other contributions seem to employ manipulations or strategies other than (and sometimes along with) those identified here;[8] regrettably, this interesting work does not fall entirely within the scope of this effort.

Another group of papers not dealt with in the present study are those dealing with an infinite number of variables or constraints, although a number of contributions along these lines have been made, particularly in the linear case—see, e.g., [Charnes, Cooper and Kortanek 69], [Hopkins 69]. Nor do we consider the literature on mathematical programs in continuous time (a recent contribution with a good bibliography is [Grinold 68]), or literature on the interface between mathematical programming and optimal control theory (e.g., [Dantzig 66], [Rosen 67], [Van Slyke 68]).

1.3 *Notation*

Although the notation we employ is not at odds with customary usage, the reader should keep a few conventions in mind.

Lowercase letters are used for scalars, scalar-valued functions, and vectors of variables or constants. Except for gradients (e.g., $\nabla f(x) = (\partial f(x)/\partial x_1, \cdots, \partial f(x)/\partial x_n)$),

---

[7] E.g.: *Inner Linearization/Relaxation:* Abadie and Williams 63, Whinston 64. *Dualization, Outer Linearization/Relaxation:* Takahashi 64 ("global" approach), Geoffrion 68b (§6), Fox 70. *Inner Linearization, Projection, Outer Linearization/Relaxation:* Metz, Howard and Williamson 66. *Dualization/Relaxation:* Webber and White 68.

[8] E.g.: Balas 65 and 66, Bell 66, Charnes and Cooper 55, Gomory and Hu 62 (Secs. 1 and 2), Kornai and Liptak 65, Kronsjö 68, Orchard-Hays 68 (Ch. 12), Rech 66, Ritter 67b.

all vectors are column vectors unless transposed. Capital letters are used for matrices $(A, B, \text{etc.})$, sets $(X, Y, \text{etc.})$ and vector-valued functions (e.g., $G(x) = [g_1(x), \cdots, g_m(x)]^t$). The dimension of a matrix or vector-valued function is left unspecified when it is immaterial to the discussion or obvious from context. The dimension of x, however, will always be n. The symbol "$\leqq$" is used for vector inequalities, and "$\leq$" for scalar inequalities. "$\triangleq$" means "equal by definition to." The notation $s.t.$, used in stating a constrained optimization problem, means "subject to." *Convex polytope* refers to the solution set of a finite system of linear equations or inequations; it need not be a bounded set.

2. Problem Manipulations: Source of "Master" Problems

A *problem manipulation* is defined to be the restatement of a given problem in an alternative form that is essentially equivalent but more amenable to solution. Nearly all of the so-called *master* problems found in the large-scale programming literature are obtained in this way.

A very simple example of a problem manipulation is the introduction of slack variables in linear programming to convert linear inequality constraints into linear equalities. Another is the restatement of a totally separable problem like (here x_i may be a vector)

$$\text{Minimize}_{x_1, \cdots, x_k} \quad \sum_{i=1}^{k} f_i(x_i) \quad \text{s.t.} \quad G_i(x_i) \geqq 0, \quad i = 1, \cdots, k$$

as k independent problems, each of the form

$$\text{Minimize}_{x_i} \quad f_i(x_i) \quad \text{s.t.} \quad G_i(x_i) \geqq 0.$$

This manipulation crops up frequently in large-scale optimization, and will be called *separation*.

These examples, although mathematically trivial, do illustrate the customary purpose of problem manipulation: to permit existing optimization algorithms to be applied where they otherwise could not, or to take advantage in some way of the special structure of a particular problem. The first example permits the classical Simplex Method which deals directly only with equality constraints, to be applied to linear programs with inequality constraints. The second example enables solving a totally separable problem by the simultaneous solution of smaller problems. Even if the smaller problems are solved sequentially rather than simultaneously, a net advantage is still probable since for most solution methods the amount of work required increases much faster than linearly with problem size.

More specifically, the three main objectives of problem manipulation in large-scale programming seem to be:
  (a) to isolate familiar special structures imbedded in a given problem (so that known efficient algorithms appropriate to these structures can be used);
  (b) to induce linearity in a partly nonlinear problem via judicious approximation (so that the powerful linear programming algorithms can be used);
  (c) to induce separation.
We shall discuss in detail three potent devices frequently used in pursuit of these objectives: *Projection, Inner Linearization,* and *Outer Linearization.*

Projection (§2.1), sometimes known as "partitioning" or "parameterization," is a device which takes advantage in certain problems of the relative simplicity resulting when certain variables are temporarily fixed in value. In [Benders 62] it is used for objective (a) above to isolate the linear part of a "semilinear" program (see §4.1), while in [Rosen 64] it is used to induce separation (see §4.2).

Inner Linearization (§2.2) and Outer Linearization (§2.3) are devices for objective (b) long used in nonlinear programming. Inner Linearization goes back at least to [Charnes and Lemke 54], in which a convex function of one variable is approximated by a piecewise-linear convex function. Outer Linearization involves tangential approximation to convex functions as in [Kelley 60] (see §3.3). Both devices have important uses in large-scale programming. Inner Linearization is the primary problem manipulation used in the famous Dantzig-Wolfe decomposition method of linear and nonlinear programming (§4.3). One important use of Outer Linearization is as a means of dealing with nonlinearities introduced by Projection (§4.1).

Perhaps the most conspicuous problem manipulation not discussed here is *Dualization*. Long familiar in the context of linear programs, dualization of nonlinear programs[9] is especially valuable in pursuit of objectives (a) and (c). This significant omission is made because of space considerations, and also to keep the presentation as elementary as possible. One algorithm relying on nonlinear dualization is mentioned in §4.5; see also part E of Table 2 and [Geoffrion 68b; §6.1].

Other problem manipulations not discussed here, mostly quite specialized, can be found playing conspicuous roles in [Charnes and Cooper 55], [El Agizy 67], [Gomory and Hu 62], [Weil and Kettler 68].

We now proceed to discuss Projection and Inner and Outer Linearization. §3 will discuss the solution strategies that can be applied subsequent to these and other problem manipulations. The distinction between problem manipulations and solution strategies is that the former replaces an optimization problem by one that is essentially equivalent to it, while the latter replaces a problem by a sequence of related but much simpler optimization problems.

2.1 *Projection*

The problem

$$(2.1) \qquad \text{Maximize}_{x \in X; y \in Y} \quad f(x, y) \quad \text{s.t.} \quad G(x, y) \geqq 0$$

involves optimization over the joint space of the x and y variables. We define its *projection* onto the space of the y variables alone as

$$(2.2) \qquad \text{Maximize}_{y \in Y} \quad [\text{Sup}_{x \in X} f(x, y) \quad \text{s.t.} \quad G(x, y) \geqq 0].$$

The maximand of (2.2) is the entire bracketed quantity—call it $v(y)$—which is evaluated, for fixed y, as the supremal value of an "inner" maximization problem in the variables x. We define $v(y)$ to be $-\infty$ if the inner problem is infeasible. The only constraint on y in (2.2) is that it must be in Y, but obviously to be a candidate for the optimal solution y must also be such that the inner problem is feasible, i.e., y must be in the effective domain V of v, where

$$(2.3) \qquad V \triangleq \{y : v(y) > -\infty\} \equiv \{y : G(x, y) \geqq 0 \text{ for some } x \in X\}.$$

Thus we may rewrite (2.2) as

$$(2.4) \qquad \text{Maximize}_{y \in Y \cap V} \, v(y).$$

The set V can be thought of as the projection of the constraints $x \in X$

---

[9] See, e.g., Rockafellar 68, Geoffrion 69.

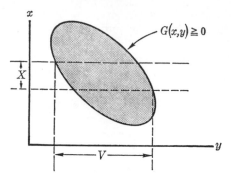

FIGURE 1. Depiction of the set V

and $G(x, y) \geq 0$ onto the space of the y variables alone. It is depicted for a simple case in Figure 1; X is an interval, the set $\{(x, y) : G(x, y) \geq 0\}$ is shaded, and the resulting V is an interval. It is often possible to obtain a more conventional and tractable representation of V than the definitional one (see, for example, the inequalities (4.5) of §4.1, [Kohler 67] and Theorem 2 of [Geoffrion 70]).

The relationship between the original problem (2.1) and its projection (2.4) is as follows.[10] The proof is elementary.

THEOREM 1. *Problem (2.1) is infeasible or has unbounded value if and only if the same is true of problem (2.4). If (x^0, y^0) is optimal in (2.1), then y^0 must be optimal in (2.4). If y^0 is optimal in (2.4) and x^0 achieves the supremum of $f(x, y^0)$ subject to $x \in X$ and $G(x, y^0) \geq 0$, then x^0 together with y^0 is optimal in (2.1). If y^0 is ϵ_1-optimal in (2.4) and x^0 is within ϵ_2 of achieving $v(y^0)$, then (x^0, y^0) is $(\epsilon_1 + \epsilon_2)$-optimal in (2.1).*

It should be emphasized that Projection is a very general manipulation—no special assumptions on X, Y, f, or G are required for Theorem 1 to hold, and any subset of variables whatsoever can be designated to play the role of y. When convexity assumptions do hold, however, the following theorem shows that (2.2) is a concave program.

THEOREM 2. *Assume that X and Y are convex sets, and that f and each component of G are concave on $X \times Y$. Then the maximand $v(y)$ in (2.2) is concave on Y.*

PROOF. Fix $y^0, y' \in Y$ and $0 < \theta < 1$ arbitrarily. Then

$$v(\theta y^0 + (1 - \theta)y') = \mathrm{Sup}_{x^0, x' \in X}\, f(\theta x^0 + (1 - \theta)x',\ \theta y^0 + (1 - \theta)y')$$

$$\text{s.t. } G(\theta x^0 + (1 - \theta)x',\ \theta y^0 + (1 - \theta)y') \geq 0$$

$$\geq \mathrm{Sup}_{x^0, x' \in X}\, f(\theta x^0 + (1 - \theta)x',\ \theta y^0 + (1 - \theta)y')$$

$$\text{s.t. } G(x^0, y^0) \geq 0,\ G(x', y') \geq 0$$

$$\geq \mathrm{Sup}_{x^0, x' \in X}\, \theta f(x^0, y^0) + (1 - \theta)f(x', y')$$

$$\text{s.t. } G(x^0, y^0) \geq 0,\ G(x', y') \geq 0$$

$$= \theta v(y^0) + (1 - \theta)v(y'),$$

----

[10] One may read (2.2) for (2.4) in Theorem 1, except that (2.2) can be feasible with value $-\infty$ when (2.1) is infeasible.

where the equality or inequality relations follow, respectively, from the convexity of X, the concavity of G, the concavity of f, and separability in x^0 and x'.

Since V is easily shown to be a convex set when v is concave, it follows under the hypotheses of Theorem 2 that (2.4) is also a concave program.

Projection is likely to be a useful manipulation when a problem is significantly simplified by temporarily fixing the values of certain variables. In [Benders 62], (2.1) is a linear program for fixed y (see §4.1). In [Rosen 64], (2.1) is a separable linear program for fixed y (cf. §4.2). See Table 2 for numerous other instances in which Projection plays an important role.

It is interesting to note that Projection can be applied sequentially by first projecting onto a subset of the variables, then onto a subset of these, and so on. The result is a dynamic-programming-like reformulation [Bellman 57], [Dantzig 59, p. 61 ff.], [Nemhauser 64]. Many dynamic programming problems can fruitfully be viewed in terms of sequential projection, and conversely, but we shall not pursue this matter here.

It may seem that the maximand of the projected problem (2.2) is excessively burdensome to deal with. And indeed it may be, but the solution strategies of §3 enable many applications of Projection to be accomplished successfully. The key strategies seem to be Relaxation preceded by Outer Linearization (cf. §4.1), the Piecewise strategy (cf. §4.2), and Feasible Directions (cf. §4.4). Of course if y is only one-dimensional, (2.2) can be solved in a parametric fashion [Joksch 64], Ritter [67a].

2.2 *Inner Linearization*

Inner Linearization is an approximation applying both to convex or concave functions and to convex sets. It is conservative in that it does not underestimate (overestimate) the value of a convex (concave) function, or include any points outside of an approximated convex set.

An example of Inner Linearization applied to a convex set X in two dimensions is given in Figure 2, where X has been approximated by the convex hull of the points $x^1, \cdots, x^5$ lying within it. X has been linearized in the sense that the approximating set is a convex polytope (which, of course, can be specified by a finite number of linear inequalities). The points $x^1, \cdots, x^5$ are called the *base*. The accuracy of the approximation can be made as great as desired by making the density of the base sufficiently high.

An example of Inner Linearization applied to a function of one variable is given in Figure 3, where the function f has been approximated on the interval $[x^1, x^5]$ by a piecewise-linear function (represented by the dotted line) that accomplishes linear interpolation between the values of f at the base points $x^1, \cdots, x^5$. The approximation

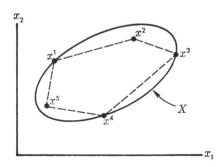

FIGURE 2. Inner Linearization of a convex set

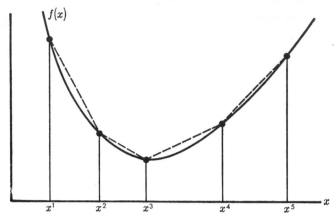

FIGURE 3. Inner Linearization of a convex function

is "inner" in the sense that the epigraph of the approximating function lies entirely within the epigraph of the approximated function. (The *epigraph* of a convex (concave) function is the set of all points lying on or above (below) the graph of the function.)

Let us further examine these two graphical examples of Inner Linearization in the context of the special problem

$$(2.5) \qquad \text{Minimize}_{x \in X} f(x) \quad \text{s.t.} \quad G(x) \leqq 0,$$

where $n = 2$, X is a convex set, and all functions are convex. Inner-linearizing X as in Figure 2 yields the approximation

$$(2.6) \quad \text{Minimize}_{\alpha \geqq 0} f(\textstyle\sum_{j=1}^{5} \alpha^j x^j) \quad \text{s.t.} \quad G(\textstyle\sum_{j=1}^{5} \alpha^j x^j) \leqq 0, \quad \textstyle\sum_{j=1}^{5} \alpha^j = 1.$$

Note that the x variables are replaced by the "weighting" variables α^j, one for each chosen base point in X. Inner-linearizing f now as in the two-dimensional analog of Figure 3 yields

$$(2.7) \quad \text{Minimize}_{\alpha \geqq 0} \textstyle\sum_{j=1}^{5} \alpha^j f(x^j) \quad \text{s.t.} \quad G(\textstyle\sum_{j=1}^{5} \alpha^j x^j) \leqq 0, \quad \textstyle\sum_{j=1}^{5} \alpha^j = 1.$$

We have taken the bases for the approximations to X and f to coincide, since normally only one base is introduced for a given problem. An exception to this general rule may occur, however, when some of the functions are separable, for then it may be desirable to introduce different bases for different subsets of variables. Suppose, for example, that $f(x) = f_1(x_1) + f_2(x_2)$, $X = R^2$, and that we wish to use $\langle x_1^1, \cdots, x_1^4 \rangle$ as a base for inner-linearizing f_1 and $\langle x_2^1, \cdots, x_2^6 \rangle$ as a base for f_2. Then the corresponding approximation to (2.5) would be

$$(2.8) \quad \begin{aligned} &\text{Minimize}_{\alpha_1 \geqq 0; \alpha_2 \geqq 0} \textstyle\sum_{j=1}^{4} \alpha_1^j f_1(x_1^j) + \textstyle\sum_{j=1}^{6} \alpha_2^j f_2(x_2^j) \\ &\text{s.t.} \quad G(\textstyle\sum_{j=1}^{4} \alpha_1^j x_1^j, \textstyle\sum_{j=1}^{6} \alpha_2^j x_2^j) \leqq 0, \quad \textstyle\sum_{j=1}^{4} \alpha_1^j = 1 \quad \text{and} \quad \textstyle\sum_{j=1}^{6} \alpha_2^j = 1. \end{aligned}$$

Problems (2.6), (2.7) and (2.8) are all convex programs.

The general nature of Inner Linearization should be clear from these examples. It is important to appreciate that there is a great deal of flexibility in applying Inner Linearization—both as to which sets and functions are inner-linearized, and as to which base is used. Inner-linearizing everything results, of course, in a linear program, although it is by no means necessary to inner-linearize everything (see §4.3). The base can be chosen to approximate the set of points satisfying any subset whatever of the given constraints; the constraints in the selected subset are replaced by the simple non-negativity conditions on the weighting variables plus the normalization constraint, while the remaining constraints are candidates for functional Inner Linearization with respect to the chosen base. Or, if desired, the base can be chosen freely from the whole space of the decision variables (this can be thought of as corresponding to the selection of an empty set of constraints). Each of the given constraints, then, is placed in one of three categories, any of which may be empty: the constraints defining the convex set approximated by the chosen base, those that are inner-linearized over the base, and all others.

Inner Linearization has long been used for convex (or concave) functions of a single variable [Charnes and Lemke 54]. It has also been used for non-convex functions of a single variable [Miller 63]. Techniques based on this manipulation are sometimes called "separable programming" methods because they deal with functions that are linearly separable into functions of one variable (e.g., $f(x) \triangleq \sum_{i=1}^{n} f_i(x_i)$).

It is easy to determine—perhaps graphically—an explicit base yielding as accurate an inner-linearization as desired for a given function of one variable. It is much more difficult, however, to do this for functions of many variables. Even if a satisfactory base could be determined, it would almost certainly contain a large number of points. This suggests the desirability of having a way to generate base points as actually needed in the course of computationally solving the inner-linearized problem. Hopefully it should be necessary to generate only a small portion of the entire base, with many of the generated points tending to cluster about the true optimal solution. Indeed there is a way to do this based on the solution strategy we call Restriction (§3.2). The net effect is that the Inner Linearization manipulation need only be done implicitly! Dantzig and Wolfe were the originators of this exceedingly clever approach to nonlinear programming [Dantzig 63a, Chapter 24]; we shall review this development in §4.3.

An important special case in which Inner Linearization can be used very elegantly concerns convex polytopes (the polytope could be the epigraph of a piecewise-linear convex function). Inner Linearization introduces no error at all in this case if the base is taken to coincide with the extreme points.[11] As above, the extreme points can be generated as needed if the implicitly inner-linearized problem is solved by Restriction. This is the idea behind the famous Decomposition Principle for linear programming [Dantzig and Wolfe 60], which is reviewed in §4.3.

For ease of reference in the sequel, the well-known theorem asserting the exactness of Inner Linearization for convex polytopes [Goldman 56], is recited here.

THEOREM 3. *Any nonempty convex polytope* $X \triangleq \{x : Ax \leq b\}$ *can be expressed as the vector sum* $\mathcal{P} + \mathcal{C}$ *of a bounded convex polyhedron* $\mathcal{P}$ *and a cone* $\mathcal{C} \triangleq \{x : Ax \leq 0\}$. $\mathcal{P}$ *in turn can be expressed as the convex hull of its extreme vectors* $\langle y_1, \cdots, y_p \rangle$, *and* $\mathcal{C}$ *can be expressed as the nonnegative linear combinations of a finite set of spanning vectors* $\langle z_1, \cdots, z_q \rangle$. (*If* $\mathcal{P}$ (*respectively* $\mathcal{C}$) *consists of just the 0-vector, take* p (*respectively* q) *equal to 0.*) *Thus there exist vectors* $\langle y_1, \cdots, y_p; z_1, \cdots, z_q \rangle$ *such that* $x \in X$ *if and only if*

$$x = \sum_{i=1}^{p} \alpha_i y_i + \sum_{i=1}^{q} \beta_i z_i$$

for some nonnegative scalars $\alpha_1, \cdots, \alpha_p, \beta_1, \cdots, \beta_q$ *such that* $\sum_{i=1}^{p} \alpha_i = 1$. *Moreover, if the rank of A equals n (the number of its columns), then a representation with a minimal*

---

[11] It is also necessary, of course, to introduce the extreme rays if the polytope is unbounded.

number of vectors is obtained by letting the y_i's be the extreme vectors of X and by letting the z_i's be distinct nonzero vectors in each of the extreme rays of $\mathcal{C}$; this minimal representation is unique up to positive multiples of the z_i's.

It should be noted that in mathematical programming the rank of A usually equals n, since nonnegativity constraints on the variables are usually included in X. If this is not the case, then X can always be imbedded in the nonnegative orthant of R^{n+1} by a simple linear transformation (viz., put $x_i = y_i - y_o$, where $y_i \geq 0$, $i = 0, \cdots, n$).

There are also results having to do with economical inner linearizations of nonpolyhedral sets. For example, there is the Theorem of Krein and Milman [Berge 63, p. 167] that every closed, bounded, nonempty convex set is the convex hull of its extreme points. Usually, however, it suffices to know that a convex set or function can be represented as accurately as desired by Inner Linearization over a sufficiently dense base.

2.3 *Outer Linearization*

Outer Linearization is complementary in nature to Inner Linearization, and also applies both to convex (or concave) functions and to convex sets.

An example as applied to a convex set in two dimensions is given by Figure 4, where X has been approximated by a containing convex polytope that is the intersection of the containing half-spaces $H_1, \cdots, H_4$. The first three are actually supporting half-spaces that pass, respectively, through the points x^1, x^2, and x^3 on the boundary of X.

An example as applied to a function of one variable is given in Figure 5, where the function f has been approximated by the piecewise-linear function that is the upper envelope, or pointwise maximum, of the linear supporting functions $s_1(x), \cdots, s_5(x)$ associated with the points $x^1, \cdots, x^5$. A *linear support* for a convex function f at the point $\bar{x}$ is defined as a linear function with the property that it nowhere exceeds f in value, and equals f in value at $\bar{x}$.[12] The epigraph of the approximating function contains the epigraph of the approximated function when Outer Linearization is used.

Obviously Outer Linearization is opposite to Inner Linearization in that it generally underestimates (overestimates) the value of a convex (concave) function, and includes not only the given convex set but points outside as well. The notion of conjugacy (see, e.g., [Rockafellar 68]) is a logical extension, but need not be pursued here.

That Outer Linearization truly linearizes a convex program like

$$(2.9) \qquad \text{Minimize}_{x \in X} f(x) \quad \text{s.t.} \quad G(x) \leq 0,$$

should be clear. The approximation of X by a containing convex polytope can only introduce linear constraints; the approximation of g_i by the pointwise maximum of a collection of p_i linear supports, say, obviously leads to p_i linear inequalities; and the approximation of f by the pointwise maximum of p linear supports leads to p additional linear inequalities after one invokes the elementary manipulation of minimizing an upper bound on f in place of f itself.[13] If *all* nonlinear functions are dealt with in this fashion, the approximation to (2.9) is a linear program.

As with Inner Linearization, there is great latitude concerning which sets and func-

---

[12] If f is differentiable at $\bar{x}$, then $f(\bar{x}) + \nabla f(\bar{x})(x - \bar{x})$ is a linear support at $\bar{x}$.
[13] E.g., $\text{Min}_{x \in X} \text{Max}_i \{s_i(x)\} = \text{Min}_{x \in X; \sigma} \sigma$ s.t. $\sigma \geq s_i(x)$, all i.

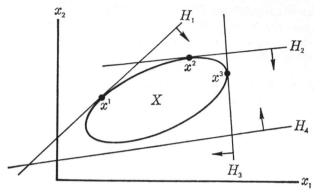

FIGURE 4. Outer Linearization of a convex set

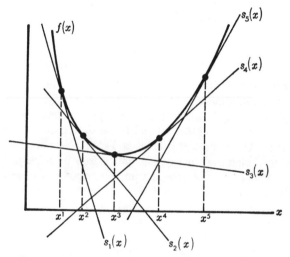

FIGURE 5. Outer Linearization of a convex function

tions are to be outer-linearized, and which approximants[14] are to be used. In general, the objective function may or may not be outer-linearized, and each constraint is placed into one of three categories: the ones that together define a convex set to be outer-linearized, the ones that are outer-linearized individually, and the ones that are not outer-linearized at all.

The main obstacle faced with Outer Linearization is that an excessive number of approximants may be required for an adequate approximation, especially for sets in more than two dimensions and functions of more than one variable. Fortunately it turns out that it is usually possible to circumvent this difficulty, for there is a solution strategy applicable to the outer-linearized problem that enables approximants to be generated economically as needed without having to specify them in advance. We call this strategy Relaxation. The net effect is that the Outer Linearization manipulation need only be done implicitly. Two pioneering papers on this approach to nonlinear

---

[14] For the sake of unified terminology, we use the term *approximant* for a containing or supporting half-space of a convex set, and also for a linear bounding function or linear support of a convex function.

programming are [Kelley 60] and [Dantzig and Madansky 61]. Relaxation and the first of these papers are discussed in §3.3.

In large-scale programming, Outer Linearization is especially important in conjunction with Projection and Dualization. See, for example, the discussion of [Benders 62] in §4.1.

Approximation by Outer Linearization naturally raises the question of the existence of a supporting approximant at a given point. The main known result along these lines is that every boundary point of a convex set in R^n must have at least one supporting half-space passing through it. It follows that every closed convex set can be represented as the intersection of its supporting half-spaces [Berge 63, p. 166].[15] It also follows that every convex (or concave) function with a closed epigraph has a supporting half-space to its epigraph at every point where the function is finite. Unfortunately, this is not quite the same as the existence of a linear support at every such point, since the supporting half-space may be "vertical" when viewed as in Figure 5. Various mild conditions could be imposed to preclude this kind of exceptional behavior, but for most purposes one may avoid the difficulty by simply working directly with the epigraph of a convex function.

3. Solution Strategies: Source of "Subproblems"

The previous section described several prominent problem manipulations for restating a given problem in a more or less equivalent form. The result is often referred to in specific applications as a "master" problem. Typically one then applies a solution strategy designed to facilitate optimization by reduction to a sequence of simpler optimization problems. Quite often this leads to *subproblems* amenable to solution by specialized algorithms. There are perhaps a half dozen principal solution strategies, each applicable to a variety of problems and implementable in a variety of ways. This section presents three such strategies that seem to be especially useful for large-scale problems: the so-called *Piecewise*, *Restriction* and *Relaxation* strategies. See Table 2 for a classification of many known algorithms in terms of the solution strategy they can be viewed as using.

The Piecewise strategy is appropriate for problems that are significantly simpler if their variables are temporarily restricted to certain regions of their domain. The domain is (implicitly) sub-divided into such regions, and the problem is solved by considering the regions one at a time. Usually it is necessary to consider only a small fraction of all possible regions explicitly. The development of the Piecewise strategy for large-scale programming is largely due to J. B. Rosen, whose various Partition Programming algorithms invoke it subsequent to the Projection manipulation.

Restriction is often appropriate for problems with a large number of nonnegative variables. It enables reduction to a sequence of problems in which most of the variables are fixed at zero. The Simplex Method itself turns out to be a special form of Restriction for linear programming, although the strategy also applies to nonlinear problems. Restriction is almost always used if Inner Linearization has been applied.

Relaxation is useful for problems with many inequality constraints. It reduces such a problem to a recursive sequence of problems in which many of these constraints are ignored. The Dual Method of linear programming is a special form of Relaxation, although the strategy applies equally well to nonlinear problems. Outer Linearization is almost always followed by Relaxation.

---

[15] Of course, a convex polytope by definition admits an exact outer-linearization using only a *finite* number of approximants.

Perhaps the most important solution strategy *not* discussed here is the well-known *Feasible Direction* strategy [Zoutendijk 60], which reduces a problem with differentiable functions to a sequence of one-dimensional optimization problems along carefully chosen directions. Most of the more powerful primal nonlinear programming algorithms utilize this strategy, but their application to large-scale problems is frequently hampered by non-differentiability (if Dualization or Projection is used) if not by sheer size (especially if Inner or Outer Linearization is used). See §4.4 for an instance in which the first obstacle can be surmounted.

We have also omitted discussion of the *Penalty* strategy (e.g., [Fiacco and Mc-Cormick 68]), which reduces a constrained problem to a sequence of essentially unconstrained problems via penalty functions. The relevance of this strategy to large-scale programming is hampered by the fact that penalty functions tend to destroy linearity and linear separability.

3.1 *Piecewise Strategy*

Suppose that one must solve

$$(3.1) \qquad \text{Maximize}_{y \in Y}\, v(y),$$

where v is a "piecewise-simple" function (e.g., piecewise-linear or piecewise-quadratic) in the sense that there are regions (pieces) P^1, P^2, $\cdots$ of its domain such that v coincides with a relatively tractable function v^k on P^k. The situation can be depicted as in Figure 6, in which Y is a disk partitioned into four regions. Let us further suppose that v is concave on the convex set Y and that, given any particular point in Y, we can explicitly characterize the particular piece to which that point belongs, as well as v on that piece. Then it is natural to consider solving (3.1) in the following piecemeal fashion that takes advantage of the piecewise-simplicity of v. Note that it is unnecessary to explicitly characterize all of the pieces in advance.

The Piecewise Strategy

Step 1 Let a point y^0 feasible in (3.1) be given. Determine the corresponding piece P^0 containing y^0 and the corresponding function v^0.

Step 2 Maximize $v^0(y)$ subject to $y \in Y \cap P^0$. Let y' be an optimal solution (an infinite optimal value implies termination).

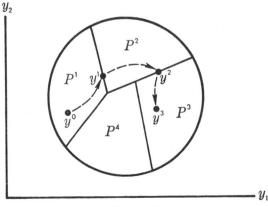

FIGURE 6.

Step 3 Determine a piece P' adjacent to P^0 at y' such that $v(y) > v(y')$ for some $y \epsilon Y \cap P'$ [if none exists, y' is optimal in (3.1)]. Determine the corresponding function v' and return to Step 2 with P', v', and y' in place of P^0, v^0, and y^0.

A hypothetical trajectory for y is traced in Figure 6 as a dotted line. Optimizations (Step 2) were performed in three regions before the optimal solution of (3.1) was found.

The problem at Step 2 has a simpler criterion function than (3.1) itself, although it has more constraints $(y \in P^0)$. If it is sufficiently simple by comparison with (3.1), then the Piecewise strategy is likely to be advantageous provided Steps 1 and 3 are not too difficult. Both Steps 2 and 3 can give rise to "subproblems" when this strategy is used for large-scale programming.

The principal use of the Piecewise strategy in large-scale programming is for problems resulting from Projection or Dualization. In both cases [cf. (2.2)], v involves the optimal value of an associated "inner" optimization problem parameterized by y. Evaluating v requires solving the inner problem, and so v is not explicitly available in closed form. Fortunately, it usually happens that evaluating $v(y^0)$ yields as a by-product a characterization of the piece P^0 containing y^0 on which v has relatively simple form. We shall illustrate this with a simple example. See also §4.2 and [Geoffrion 68b; §5].

The Piecewise strategy can also be used to motivate a generalization of the Simplex Method that allows the minimand to be a sum of piecewise-linear univariate convex functions [Orden and Nalbandian 68].

Example

Constrained games and similar applications can lead to problems of the form

$$(3.2) \qquad \text{Maximize}_{y \in Y} [\text{Minimum}_{x \geq 0} \{ H^t(y)x \quad \text{s.t.} \quad Ax = b \}],$$

where $H(\cdot)$ is a concave vector-valued function on the convex set Y. The maximand of (3.2), v, is concave because it is the pointwise minimum of a collection of concave functions of y. Suppose that we evaluate v at $y^0 \in Y$, with the corresponding optimal solution of the inner problem being x^0. The value is $H^t(y^0)x^0$. We know from the elementary theory of linear programming that, since changes in y cannot affect the feasibility of x^0, x^0 remains an optimal solution of the inner problem as y varies so long as the "reduced costs" remain of the right sign. Hence the value of $v(y)$ is $H^t(y)x^0$ for all y such that

$$(3.3) \qquad (H^B(y))^t B^{-1} A_{\cdot j} - h_j(y) \leq 0, \quad \text{all nonbasic } j,$$

where $A_{\cdot j}$ is the jth column of A, and the component functions of H^B correspond to the variables x_i in the optimal basis matrix B at y^0. Thus we see how to accomplish Step 1, and the problem to be solved at Step 2 is

$$(3.4) \qquad \text{Maximize}_{y \in Y} H^t(y)x^0 \quad \text{s.t.} \quad (3.3).$$

Note that (3.4) has the advantage over (3.2) of an explicit criterion function. Since $x^0 \geq 0$, $H^t(\cdot)x^0$ is concave on Y.

Suppose that y' is an optimal solution of (3.4).[16] If y' is not optimal in (3.2), then there must be an alternate optimal basis B' at y' such that the corresponding problem

---

[16] It may be difficult to find a global optimum of (3.4) if H is not linear, for then (3.3) need not define a convex feasible region (unless $B^{-1}A_{\cdot j} \leq 0$ for all nonbasic j). Fortunately, however, it can be seen from the concavity of v that a local optimum will generally suffice, although finite termination may now be in jeopardy.

(3.4) admits an improved solution. At worst, such an "improving" basis could be found by enumerating the alternative optimal bases at y'. At best, an improving basis would be revealed by a single active constraint among those of (3.3) at y'. One could also compute an improving feasible direction z' for (3.2) at y' (cf. §4.4); the appropriate improving basis would then be revealed by a parametric linear programming analysis of the inner problem.

3.2 *Restriction*

Restriction is a solution strategy principally useful for problems with many non-negative variables, the data associated with some of which perhaps being only implicitly available. Combinatorial models and Inner Linearization are two fertile sources of such problems.

The basic idea is as follows: solve the given problem subject to the additional restriction that a certain subset of the variables must have value 0; if the resulting solution does not satisfy the optimality conditions of the given problem, then "release" one or more restricted variables (allow them to be nonnegative) and solve this less-restricted problem; continue in this fashion until the optimality conditions of the given problem are satisfied, at which point the procedure terminates. An important refinement forming an integral part of the strategy involves *adding* variables to, as well as releasing them from, the restricted set. Note that the variables restricted to 0 essentially drop out of the problem, thereby reducing its size and avoiding the need for knowing the associated data explicitly. If (as is usually the case) only a fairly small proportion of all variables actually are active (positive) at an optimal solution, then this strategy becomes quite attractive.

The earliest and most significant embodiment of the Restriction strategy turns out to be the Simplex Method for linear programming itself. It can be shown, as we shall indicate, that a natural specialization of Restriction to the completely linear case yields the very same sequence of trial solutions as does the ordinary Simplex Method. All of the "column-generation" schemes for implementing the Simplex Method for linear programs with a vast number of variables can therefore be viewed in terms of Restriction. We shall review one of these schemes [Gilmore and Gomory 61] at the end of this section.[17] The usefulness of Restriction is not, however, limited to the domain of linear programming. It will be shown in §4.3 how this strategy yields, in a nonlinear case, variations of the Dantzig-Wolfe method for convex programming.

Formal Statement. Consider the problem

(3.5) $$\text{Maximize}_{x \in X} f(x) \quad \text{s.t.} \quad g_i(x) \geq 0, i = 1, \cdots, m,$$

where f is a concave function on the nonempty convex set $X \subseteq R^n$ and the functions $g_1, \cdots, g_m$ are all linear. All nonlinear constraints, as well as any linear constraints that are not to be restricted, are presumed to be incorporated in X. The typical restricted version of (3.5) is the (still concave) problem

(3.6) $$\text{Maximize}_{x \in X} \quad f(x) \quad \text{s.t.} \quad g_i(x) = 0, i \in S,$$
$$g_i(x) \geq 0, i \notin S,$$

where S is a subset of the m constraint indices. [Note that we are presenting Restriction in a seemingly more general setting than the motivational one above in that general linear inequality constraints, as well as simple variable nonnegativities, are allowed to

---

[17] Another column-generating scheme is explained in §4.3. See also part B of Table 1.

be restricted to equality. Actually, the present setting is no more general since slack variables could be introduced to accommodate the restriction of general linear inequalities.] Some, none, or all of the $x_j \geq 0$ type constraints (if any) may be included among $g_1, \cdots, g_m$. The analyst is free to choose the linear inequality constraints to associate with X; the rest are candidates for restriction.

An optimal solution of the restricted problem (3.6) will be denoted by x^s, and a corresponding optimal multiplier vector (which, under mild assumptions, must exist) is denoted by $\mu^s = (\mu_1^s, \cdots, \mu_m^s)$. The pair (x^s, μ^s) satisfies the Kuhn-Tucker optimality conditions for (3.6), namely

  (i) x^s maximizes $f(x) + \sum_{i=1}^m \mu_i^s g_i(x)$ over X
  (ii) x^s is feasible in (3.6)
  (iii) $\mu_i^s \geq 0$, $i \notin S$
  (iv) $\mu_i^s g_i(x^s) = 0$, $i \notin S$.

We are now ready to give a formal statement of Restriction applied to (3.5). Notice that not only are constraints released from the current restricted set S at each iteration, but additions are also made whenever $g_i(x^s) = 0$ for some $i \notin S$, provided that $f(x^s)$ has just increased.

The Restriction Strategy

Step 1 Put $\bar{f} = -\infty$ and S equal to any subset of indices such that the corresponding restricted problem (3.6) is feasible.

Step 2 Solve (3.6) for an optimal solution x^s and associated optimal multipliers μ_i^s (if it has unbounded optimal value, the same must be true of the given problem (3.5) and we terminate). If $\mu_i^s \geq 0$ for all $i \in S$, then terminate (x^s is optimal in (3.5)); otherwise, go on to Step 3.

Step 3 Put V equal to any subset of S that includes at least one constraint for which $\mu_i^s < 0$. If $f(x^s) > \bar{f}$, replace $\bar{f}$ by $f(x^s)$ and S by $E - V$, where $E \triangleq \{1 \leq i \leq m : g_i(x^s) = 0\}$; otherwise (i.e., if $f(x^s) = \bar{f}$), replace S by $S - V$. Return to Step 2.

We assume that the given problem (3.5) admits a feasible solution, so that Step 1 is possible. To ensure that Step 2 is always possible, we also assume that the restricted problem (3.6) admits an optimal solution and multiplier vector whenever it is feasible and has finite supremal value. It is a straightforward matter to show that the termination conditions of Step 2 are valid, and Step 3 is obviously always possible. Thus the strategy is well defined, although we have deliberately not specified how to carry out each step.

An important property is that the sequence $\langle f(x^s) \rangle$ is nondecreasing. Thus the strategy yields an improving sequence of feasible solutions to (3.5). Moreover, $\langle f(x^s) \rangle$ can be stationary in value at most a finite number of consecutive times, since the role of $\bar{f}$ at Step 3 is to ensure that S is augmented (before deletion by V) only when $f(x^s)$ has just increased. Hence termination must occur in a finite number of steps, for there is only a finite number of possibilities for S and each increase in $f(x^s)$ precludes repetition of any previous S.

Options and Relation to the Simplex Method. Let us now consider the main options of Restriction beyond the decision as to which of the linear inequality constraints will comprise $g_1, \cdots, g_m$.

  (i) How to select the initial S at Step 1?
  (ii) How to solve (3.6) for (x^s, μ^s) at Step 2?
  (iii) What criterion to use in selecting V at Step 3?

How these options are exercised exerts a great influence upon the efficiency.

As stated above, there is an intimate relationship between Restriction and the Simplex Method in the completely linear case. Given the linear program

$$\text{Maximize}_x \quad c^t x \quad \text{s.t.} \quad Ax = b, \quad x \geqq 0,$$

define (3.5) according to the identifications

$$f(x) = c^t x$$

$$g_i(x) = x_i, \quad \text{all } i$$

$$X = \{x : Ax = b\},$$

and specialize Restriction as follows: let the initial S be chosen to coincide with the nonbasic variables in an initial basic feasible solution, and select V at Step 3 to be the index of the most negative μ_i^s. It can then be shown, under the assumption of non-degeneracy, that Restriction is equivalent to the usual Simplex Method in that the set of nonbasic variables at the νth iteration of the Simplex Method necessarily coincides with E at the νth iteration of Restriction, and the νth basic feasible solution coincides with the νth optimal solution x^s of (3.6). Thus Restriction can be viewed as one possible strategic generalization of the Simplex Method. Not only is this an interesting fact in its own right, but it also permits us to draw some inferences—as we shall see in the discussion below—concerning how best to exercise the options of Restriction.

Step 1. The selection of the initial S should be guided by two objectives: to make the corresponding restricted problem easy to solve by comparison with the given problem, and to utilize any prior knowledge that may be available concerning which of the g_i constraints are likely to hold with equality at an optimal solution. In the Simplex Method, for example, the initial choice of S implies that the restricted problem is trivial since it has a unique feasible solution; at every subsequent execution of Step 2, the restricted problem remains nearly trivial with essentially only one free variable (the entering basic variable). Useful prior knowledge is often available if the given problem is amenable to physical or mathematical insight or if a variant has been solved previously.

Step 2. How to solve the restricted problem for (x^s, μ^s) at Step 2 depends, of course, on its structure. Hopefully, enough constraints will be restricted to equality to make it vastly simpler than the original problem. In any event, it is advisable to take advantage of the fact that a *sequence* of restricted problems must be solved as the Restriction strategy is carried out. Except for the first execution of Step 2, then, what is required is a solution *recovery* technique that effectively utilizes the previous solution. The pivot operation performs precisely this function in the Simplex Method, and serves as an ideal to be approached in nonlinear applications of Restriction.

It is worth mentioning that many solution (or solution recovery) techniques that could be used for the restricted problem automatically yield μ^s as well as x^s. When this is not the case, one may find μ^s once x^s is known by solving a linear problem if f and the constraint functions defining X are differentiable, since under these conditions the Kuhn-Tucker optimality conditions for (3.6) in differential form become linear in μ.

Step 3. Perhaps the most conspicuous criterion for choosing V at Step 3 is to let it be the index of the constraint corresponding to the most negative μ_i^s. One rationale for this criterion is as follows. Suppose that μ^s is unique. It can then be shown (see [Geoffrion 69] or [Rockafellar 68]) that the optimal value of the restricted problem is differentiable as a function of perturbations about 0 of the right-hand side of the g_i

constraints, and that $-\mu_i{}^s$ is the partial derivative of the optimal value with respect to such perturbations of the ith constraint. Thus the most negative $\mu_i{}^s$ identifies the constraint in S whose release will lead to the greatest initial rate of improvement in the value of f as this constraint is permitted to deviate positively from strict equality. It can be argued that μ^s is likely to be unique, but if we drop this supposition, then $-\mu_i{}^s$ still provides an upper bound on the initial rate of improvement even though differentiability no longer holds.

This most-negative-multiplier criterion is precisely the usual criterion used by the Simplex Method in its version of Step 3 to select the entering basic variable, but it is by no means the only criterion used. The extensive computational experience presently available with different criteria used in the Simplex Method may permit some inferences to be drawn concerning the use of the analogous criteria in the nonlinear case. It has been observed [Wolfe and Cutler 63], for example, that the most-negative-multiplier criterion typically leads to a number of iterations equal to about twice the number of constraints, and that other plausible criteria can be expected to be consistently better by no more than a factor of two or so.[18] Lest it be thought that V must necessarily be a singleton, we note that we may interpret Wolfe and Cutler to have also observed [ibid., p. 190] that choosing V to consist of the five most negative multipliers reduced the number of iterations by a factor of two as compared with the single-most-negative-multiplier choice.[19] Of course, this increases the time required to solve each restricted problem. Experience such as this should at least be a source of hypotheses to be examined in nonlinear applications of Restriction.

Mechanizing the "Pricing" Operation. Each iteration of Restriction requires determining whether there exists a negative multiplier and, if so, at least one must be found. In the ordinary Simplex Method, which as we have indicated can be viewed as a particular instance of Restriction, this was originally done enumeratively by scanning the row of reduced costs for an entry of the "wrong" sign. To deal with large numbers of variables, however, it is desirable whenever possible to replace this enumeration by an algorithm that exploits the structure of the problem. This is referred to as *mechanized pricing.*

Mechanized pricing is widely practiced in the context of linear programming, where it is often referred to as *column-generation.* Since the pioneering paper [Ford and Fulkerson 58], many authors have shown how pricing could be mechanized by means of subsidiary network flow algorithms, dynamic programming, integer programming, and even linear programming. See the references of part B of Table 1, [Balinski 64], and [Gomory 63]. It will suffice to mention here but one specific illustration: the cutting-stock problem as treated by [Gilmore and Gomory 61]. See also §4.3.

Cutting-Stock Problem. A simple version of Gilmore and Gomory's cutting-stock problem, without the integrality requirement on x, is

$$(3.7) \qquad \text{Minimize}_{x \geq 0} \sum_j x_j \quad \text{s.t.} \quad \sum_j a_{ij} x_j \geq r_i, \qquad i = 1, \cdots, m,$$

---

[18] An example of another plausible criterion is this: select V to be the index of the constraint which, when deleted from S, will result in the greatest possible improvement in the optimal value of the restricted problem. Of course, this criterion is likely to be prohibitively expensive computationally in the nonlinear case.

[19] This is known as *multiple pricing*, a feature used in most production linear programming systems designed for large-scale problems. See, for example, [Orchard-Hays 68, §6.1].

where a_{ij} is the number of pieces of length l_i produced when the cutting knives are set in the jth pattern, r_i is the minimum number of required pieces of length l_i , and x_j is the number of times a bar of stock is cut according to pattern j. The number of variables is very large because of the great variety of ways a bar of stock can be cut. It is easy to see that each column of the matrix A is of the form $(y_1, \cdots, y_m)^t$, where y is a vector of nonnegative integers satisfying $\sum_{i=1}^m l_i y_i \leq \lambda$ (λ is the length of a bar of stock); and conversely, every such vector corresponds to some column (assuming that all possible patterns are allowed). When Restriction is applied to (3.7) in the form of the Simplex Method, it follows that the problem of determining the most negative multiplier can be expressed as the subsidiary optimization problem

$$(3.8) \qquad \text{Minimize}_{y \geq 0} \quad 1 - u^t y \quad \text{s.t.} \quad \sum_{i=1}^m l_i y_i \leq \lambda, \qquad y \text{ integer,}$$

where u is a known vector of the current "simplex multipliers." If slack variables are given priority over structural variables in determining entering basic variables (cf. §4.3), then u can be assumed nonnegative and (3.8) is a problem of the well known "knapsack" variety, for which very efficient solution techniques are available. See [Gilmore and Gomory 61] for full details.

3.3 *Relaxation*

Whereas Restriction is a solution strategy principally useful for problems with a large number of variables, the complementary strategy of *Relaxation* is primarily useful for problems with a large number of inequality constraints, some of which may be only implicitly available. Such problems occur, for example, as a result of Outer Linearization.[20] One of the earliest uses of Relaxation was in [Dantzig, Fulkerson, and Johnson 54, 59], and since that time this strategy has appeared in one guise or another in the works of numerous authors.[21] We discuss [Kelley 60] at the end of this section, and [Benders 62] in §4.1.

The essential idea of Relaxation is this: solve a relaxed version of the given problem ignoring some of the inequality constraints; if the resulting solution does not satisfy all of the ignored constraints, then generate and include one or more violated constraints in the relaxed problem and solve it again; continue in this fashion until a relaxed problem solution satisfies all of the ignored constraints, at which point an optimal solution of the given problem has been found. An important refinement involves *dropping* amply satisfied constraints from the relaxed problem when this does not destroy the inherent finiteness of the procedure. We give a formal statement of Relaxation (with the refinement) below.

Relaxation and Restriction are complementary strategies in a very strong sense of the word. In linear programming, for example, whereas a natural specialization of Restriction is equivalent to the ordinary Simplex Method, it is also true [Geoffrion 68a] that a similar specialization of Relaxation is equivalent to Lemke's Dual Method.

---

[20] Relaxation can also be useful for dealing with large numbers of nonnegative variables; when a constraint such as $x_j \geq 0$ is relaxed the variable x_j can often be substituted out of the problem entirely ([Ritter 67c], [Webber and White 68]).

[21] *Relaxation* without problem manipulation is used in Dantzig 55a, Sec. 3; Stone 58; Thompson, Tonge and Zionts 66; Ritter 67c; Grigoriadis and Ritter 69, Robers and Ben-Israel 69, §3. The following papers all use the pattern *Outer Linearization/Relaxation*: Cheney and Goldstein 59; Kelley 60; Dantzig and Madansky 61, p. 174; Parikh 67; Veinott 67. The references of part A of Table 2 all use the pattern *Projection, Outer Linearization/Relaxation*. See also the second footnote in §1.2.

It follows, very significantly, that *Restriction (Relaxation) applied to a linear program essentially corresponds to Relaxation (Restriction) applied to the dual* linear program. In fact [ibid.], *the same assertion holds for quite general convex programs as well.* This complementarity makes it possible to translate most statements about Restriction into statements about Relaxation, and conversely.

Since we have already given a relatively detailed discussion of Restriction, a somewhat abbreviated discussion of Relaxation will suffice. See [ibid.] for a more complete discussion.

Formal Statement. Let $f, g_1, \cdots, g_m$ be concave functions on a nonempty convex set $X \subseteq R^n$. The concave program

$$(3.9) \qquad \text{Maximize}_{x \in X} \quad f(x) \quad \text{s.t.} \quad g_i(x) \geq 0, \qquad i = 1, \cdots, m$$

is solved by solving a sequence of relaxed problems of the form

$$(3.10) \qquad \text{Maximize}_{x \in X} \quad f(x) \quad \text{s.t.} \quad g_i(x) \geq 0, \qquad i \in S,$$

where S is a subset of $\{1, \cdots, m\}$. Assume that (3.10) admits an optimal solution x^s whenever it admits a feasible solution and its maximand is bounded above on the feasible region, and assume further that an initial subset of constraint indices is known such that (3.10) has a finite optimal solution. (This assumption can be enforced, if necessary, by enforcing continuity of all functions and compactness of X.)

Under these assumptions, it is not difficult to show that the following strategy is well defined and terminates in a finite number of steps with either an optimal solution of the given problem (3.9) or an indication that none exists; moreover, in the first case a nonincreasing sequence $\langle f(x^s) \rangle$ of upper bounds on the optimal value of (3.9) is obtained and the first solution of (3.10) that is feasible in (3.9) is also optimal. This version of Relaxation deletes amply satisfied constraints from S so long as $\langle f(x^s) \rangle$ is decreasing.

The Relaxation Strategy

Step 1 Put $\bar{f} = \infty$ and S equal to any subset of indices such that the corresponding relaxed problem (3.10) has a finite optimal solution.

Step 2 Solve (3.10) for an optimal solution x^s if one exists; if none exists (i.e., if the relaxed problem is infeasible), then terminate (the given problem is infeasible). If $g_i(x^s) \geq 0$ for all $i \notin S$, then terminate (x^s is optimal in the given problem); otherwise, go on to Step 3.

Step 3 Put V equal to any subset of constraint indices that includes at least one constraint such that $g_i(x^s) < 0$. If $f(x^s) < \bar{f}$, replace $\bar{f}$ by $f(x^s)$ and S by $E \cup V$, where $E \triangleq \{i \in S : g_i(x^s) = 0\}$; otherwise (i.e., if $f(x^s) = \bar{f}$), replace S by $S \cup V$. Return to Step 2.

Discussion

As with Restriction, the analyst has considerable leeway concerning how he applies the Relaxation strategy. For instance, he can select the constraints that are to be candidates for Relaxation $(g_1, \cdots, g_m)$ in any way he wishes; the rest comprise X. He is free to choose the initial S so as to allow an easy start, or to take advantage of prior knowledge concerning which of the constraints might be active at an optimal solution. He can choose the most effective solution recovery technique to reoptimize the successive relaxed problems. And, very importantly, he can choose the criterion by which V will be selected at Step 3 and the method by which the criterion will be implemented.

Probably the most natural criterion is to let V be the index of the most violated constraint. This is the criterion most commonly employed in the Dual Method of linear programming, for example, although other criteria are possible. The complementarity between Relaxation and Restriction mentioned earlier enables us to interpret existing computational experience in linear programming so as to shed light on the merits and demerits of several alternative criteria. The discussion of Step 3 of Restriction should make further discussion of this point unnecessary. We should remark, however, that in some applications (e.g., [Dantzig, Fulkerson and Johnson 54], [Gomory 58], [Kelley 60]) only one or a few violated constraints are accessible each time the relaxed problem is solved, and it is therefore indicated that these be used regardless of whether they satisfy any particular criterion. In other applications a criterion such as "most violated constraint" is within the realm of attainability, and can be approached via a subsidiary linear program [Benders 62], network flow problem [Gomory and Hu 62], or some other subsidiary optimization problem that is amenable to efficient solution. This is the counterpart of mechanized pricing in Restriction.

Restriction and Relaxation, opposites though they are to one another, are by no means incompatible. In fact it can be shown [Geoffrion 66 and 67] that both strategies can be used simultaneously. The reduced problems become still more manageable, but assurance of finite termination requires more intricate control.

The Cutting-Plane Method. One important use of Relaxation occurs, as we have mentioned, in connection with problems that have been outer-linearized. This will be illustrated in the simplest possible setting in terms of the problem

$$(3.11) \qquad \text{Minimize}_{x \geq 0} \quad c^t x \quad \text{s.t.} \quad Ax \leq b, \qquad g(x) \leq 0,$$

where g is a convex function that is finite-valued on

$$X \triangleq \{x \geq 0 : Ax \leq b\}.$$

If one manipulates (3.11) by invoking an arbitrarily fine outer-linearization of g and then applies the Relaxation strategy with the new approximating constraints as the candidates for being relaxed, the resulting procedure is that of [Kelley 60].

Let us assume for simplicity that g is differentiable on X.[22] Then g has a linear support $g(\bar{x}) + \nabla g(\bar{x})(x - \bar{x})$ at every point $\bar{x}$ in X, where $\nabla g(\bar{x})$ is the gradient of g at $\bar{x}$, and so (3.11) is equivalent to

$$(3.12) \quad \text{Minimize}_{x \in X} \quad c^t x \quad \text{s.t.} \quad g(\bar{x}) + \nabla g(\bar{x})(x - \bar{x}) \leq 0, \qquad \text{all } \bar{x} \in X.$$

The Relaxation strategy is the natural one for solving (3.12), since it avoids the need to determine in advance all of the linear supports of g. At each iteration, a relaxed version of this problem with a finite number of approximating constraints is solved. The optimal solution $\hat{x}$ of the relaxed problem is feasible in (3.12) if and only if $g(\hat{x}) \leq 0$; if $g(\hat{x}) > 0$, then evaluation of $\nabla g(\hat{x})$ yields a violated constraint that must be appended to the current relaxed problem. Since each relaxed problem is a linear program that will be augmented by a violated constraint, it is natural to reoptimize it using postoptimality techniques based on the Dual Method for linear programming.

It is easy to generalize this development to cover the case in which (3.11) has several (nonlinear) convex constraints and a convex minimand.

---

[22] The assumption of differentiability can be weakened, since it is only necessary for g to have a support at each point of X. And even this requirement can be weakened as implicitly suggested in the conclusion of §2.3 if (3.11) is phrased in terms of the epigraph of g.

It should be pointed out that dropping amply satisfied constraints from the re-laxed problem—a feature incorporated in our statement of Relaxation—appears to be questionable in this context since (3.12) has an infinite number of constraints. See, however, the recent work [Eaves and Zangwill 69] and [Topkis 69] for conditions under which convergence to an optimal solution of (3.11) is nevertheless assured in the limit.

We remark in passing that the approach of [Hartley and Hocking 63] for (3.11) can be viewed as Restriction applied to the dual of (3.12). Since Relaxation of (3.12) corresponds to Restriction of its dual, the two approaches are really equivalent.

ELEMENTS OF LARGE SCALE MATHEMATICAL PROGRAMMING

Part II: Synthesis of Algorithms and Bibliography*†‡§

ARTHUR M. GEOFFRION

University of California, Los Angeles

The problem manipulations and solution strategies of Part I of this paper are now further illustrated by combining them in various ways to yield several known algorithms. The main object is not an exposition of these algorithms, although this is certainly important; rather, we wish to focus on the principal *patterns* in which manipulations and strategies can be assembled. These patterns constitute the real common denominators in the literature on large-scale programming. See Table 2 in Part I.

4. Synthesizing Algorithms from Manipulations and Strategies

It is beyond the scope of this effort to exemplify all of the important patterns of manipulations and strategies. We shall limit our discussion to five key ones:

1. PROJECTION, OUTER LINEARIZATION/RELAXATION
2. PROJECTION/PIECEWISE
3. INNER LINEARIZATION/RESTRICTION
4. PROJECTION/FEASIBLE DIRECTIONS
5. DUALIZATION/FEASIBLE DIRECTIONS

The first pattern is illustrated in §4.1 by Benders' Partitioning Procedure for what might be called semilinear programs; the second is illustrated in §4.2 by Rosen's Primal Partition Programming algorithm for linear programs with block-diagonal structure; the third in §4.3 by Dantzig-Wolfe Decomposition; the fourth in §4.4 by a procedure the author recently developed for nonlinear programs with multidivisional structure; and the fifth in §4.5 by the "local" approach discussed by Takahashi for concave programs with "complicating" constraints. Another key pattern, OUTER LINEARIZATION/RELAXATION, was already illustrated in §3.3 with reference to Kelley's cutting-plane method. In addition, it is indicated in §4.2 how Rosen's algorithm can be used to illustrate the pattern DUALIZATION/PIECEWISE, and in §4.3 how Dantzig-Wolfe Decomposition can be used to illustrate DUALIZATION, OUTER LINEARIZATION/RELAXATION.

The discussion of the various algorithms is as uncluttered by detail as we have been

* Received April 1969.

† This is the tenth in a series of twelve expository papers commissioned jointly by the Office of Naval Research and the Army Research Office under contract numbers Nonr-4004(00) and DA 49-092-ARO-16, respectively.

‡ The reader is assumed to be familiar with Part I of this paper, which is immediately preceding in this issue.

§ Support for this work was provided by the Ford Foundation under a Faculty Research Fellowship, by the National Science Foundation under Grant GP-8740, and by the United States Air Force under Project RAND. It is a pleasure to acknowledge the helpful comments of A. Feinberg, B. L. Fox, and C. A. Holloway.

able to make it. There is little or no mention of how to find an initial feasible solution,[1] the details of computational organization, or questions of theoretical convergence. The reader is invited to ponder such questions in the light of the concepts and results advanced in the previous two sections, and then to consult the original papers.

4.1 [Benders 62]

One might refer to

$$(4.1) \qquad \text{Maximize}_{x \geq 0; y \in Y} \quad c^t x + f(y) \quad \text{s.t.} \quad Ax + F(y) \leq b$$

as a *semi-linear* program because it is a linear program in x when y is held fixed temporarily. The algorithm of [Benders 62] for this problem can be recovered by applying the pattern PROJECTION, OUTER LINEARIZATION/RELAXATION. Specifically, project (4.1) onto the space of the y variables, outer-linearize the resulting supremal value function in the maximand, and apply the Relaxation strategy to the new constraints arising as a consequence of Outer Linearization. Assume for simplicity that (4.1) is feasible and has finite optimal value.

Projection onto the space of the y variables yields

$$(4.2) \qquad \text{Maximize}_{y \in Y} \quad [f(y) + \text{Sup}_{x \geq 0} \ \{c^t x \quad \text{s.t.} \quad Ax \leq b - F(y)\}].$$

Note that the supremal value function appearing in the maximand corresponds to the linear program

$$(4.3) \qquad \text{Maximize}_{x \geq 0} \quad c^t x \quad \text{s.t.} \quad Ax \leq b - F(y).$$

This program is parameterized nonlinearly in the right-hand side by y, and our assumption implies that it has a finite optimum for at least one value of y. By the Dual Theorem, therefore, the dual linear program

$$(4.4) \qquad \text{Minimize}_{u \geq 0} \quad u^t(b - F(y)) \quad \text{s.t.} \quad u^t A \geq c^t$$

must be feasible (for all y). Let $\langle u^1, \cdots, u^p \rangle$ be the extreme points and $\langle u^{p+1}, \cdots, u^{p+q} \rangle$ representatives of the extreme rays of the feasible region of (4.4) (cf. Theorem 3). Again using the Dual Theorem, we see that (4.3) is feasible if and only if (4.4) has finite optimal value, that is, if and only if y satisfies the constraints

$$(4.5) \qquad (u^j)^t(b - F(y)) \geq 0, \qquad j = p + 1, \cdots, p + q.$$

Since we take the supremal value function in (4.2) to be $-\infty$ for y such that (4.3) is infeasible—see §2.1—we may append the constraints (4.5) to (4.2). Thus Projection applied to (4.1) yields (4.2) subject to the additional constraints (4.5).

Next we outer-linearize the supremal value function appearing in (4.2). It is easy to see, referring to (4.4), that its value is precisely

$$(4.6) \qquad \text{Minimum}_{1 \leq j \leq p} \ \{(u^j)^t(b - F(y))\}$$

for all y feasible in (4.2) with (4.5) appended. (Strictly speaking, it is accurate to call this Outer "Linearization" only if F is linear.) With this manipulation, (4.2) becomes

$$(4.7) \quad \text{Maximize}_{y \in Y} \quad [f(y) + \text{Minimum}_{1 \leq j \leq p} \ \{(u^j)^t(b - F(y))\}] \quad \text{s.t.} \quad (4.5)$$

---

[1] If one exists, it can usually be found by applying the algorithm itself to a suitably modified version of the given problem.

or, with the help of an elementary manipulation based on the fact that a minimum is really a greatest lower bound,

$$\text{(4.8)} \quad \text{Maximize}_{y \in Y : y_0} \quad f(y) + y_0 \quad \text{s.t.} \quad y_0 \leq (u^j)^t(b - F(y)), \quad j = 1, \cdots, p,$$
$$(u^j)^t(b - F(y)) \geq 0, \quad j = p + 1, \cdots, p + q.$$

This is the master problem to be solved.

Relaxation is a natural strategy for (4.8); it avoids having to determine in advance all of the vectors $u^j, j = 1, \cdots, p + q$. To test the feasibility of a trial solution $(\hat{y}_0, \hat{y})$, where $\hat{y} \in Y$, one solves the linear subproblem (4.4) with y equal to $\hat{y}$. If the infimal value is greater than or equal to $\hat{y}_0$, then $(\hat{y}_0, \hat{y})$ is feasible and therefore optimal in (4.8); $\hat{y}$, along with $\hat{x}$ equal to the optimal dual variables of (4.4), is an optimal solution of the given problem (4.1). If, on the other hand, the infimal value is less than $\hat{y}_0$, then a violated constraint of (4.8) is produced (some u^j with $1 \leq j \leq p$ is found if the infimal value is finite, while $p + 1 \leq j \leq p + q$ if it is $-\infty$). Of course, f, F, and Y must satisfy the obvious convexity assumptions if dropping amply satisfied constraints is to be justified. These assumptions will probably have to hold anyway if the relaxed problems based on (4.8) are to be concave programs (remember $u^j \geq 0$). There is, however, at least one other interesting case: if Y is a discrete set, say the integer points of some convex polytope, while f and F are linear, then (4.8) is a pure (except for y_0) integer linear program (see [Balinski and Wolfe 63], [Buzby, Stone and Taylor 65]).

The present development seems preferable to the original one since: (a) it justifies dropping amply satisfied constraints from successive relaxed versions of (4.8); (b) it retains $f(y)$ in its natural position in the criterion function of (4.8) (Benders' version of (4.8), which is also equivalent to (4.7), has y_0 alone as the criterion function and an added term $f(y)$ in the right-hand side of each of the first p constraints); and (c) its comparative simplicity suggests a generalization, with the help of nonlinear duality theory, permitting nonlinearities in x [Geoffrion 70].

4.2 [Rosen 64]

The algorithm of [Rosen 64] for the linear program

$$\text{(4.9)} \quad \text{Maximize}_{x,y} \quad b_0^t y + \sum_{i=1}^{l} b_i^t x_i \quad \text{s.t.} \quad x_i^t A_i + y^t D_i \leq c_i^t, \quad i = 1, \cdots, l$$

illustrates the pattern PROJECTION/PIECEWISE. Assume for simplicity that (4.9) is feasible and has finite optimal value.

Projection onto the y variables yields the master problem

$$\text{(4.10)} \quad \text{Maximize}_y \quad [b_0^t y + \sum_{i=1}^{l} \text{Sup}_{x_i} \{b_i^t x_i \quad \text{s.t.} \quad x_i^t A_i \leq c_i^t - y^t D_i\}],$$

where we have separated the supremum in the maximand (this separation is perhaps the main justification for using Projection).

The Piecewise strategy is appropriate for (4.10) because each supremal value in the maximand is piecewise-linear as a function of y. This follows from the elementary theory of linear programming, as we now explain. Let $\hat{y}$ be feasible in (4.10) in the sense that the maximand is not $-\infty$. Then each of the l linear programs appearing in the maximand must have a finite optimal value, and by the Dual Theorem this optimal value must be equal to that of the dual linear program

$$\text{(4.11)} \quad \text{Minimize}_{u_i \geq 0} \quad (c_i^t - \hat{y}^t D_i) u_i \quad \text{s.t.} \quad A_i u_i = b_i.$$

Let the vector $\hat{u}_i$ be an optimal solution of this program, and let the corresponding basis matrix be B_i. Since changes in y cannot affect the feasibility of $\hat{u}_i$, the optimal value of (4.11)—which is equal to the value of ith supremal value function of (4.10) at y—must be

$$(4.12) \qquad\qquad (c_i{}^t - y^t D_i)\hat{u}_i$$

so long as the "reduced costs" remain of the correct sign, that is, so long as y satisfies the condition

$$(4.13) \qquad (c_i{}^t - y^t D_i)^B B_i^{-1}(A_i)._j - (c_i{}^t - y^t D_i)_j \leq 0, \quad \text{all nonbasic } j,$$

where the superscript B masks all but the basic components of $(c_i{}^t - y^t D_i)$. Thus the master problem (4.10), confined to the linear "piece" containing $\hat{y}$, becomes the linear program

$$(4.14) \qquad \text{Maximize}_y \quad b_0{}^t y + \sum_{i=1}^l (c_i{}^t - y^t D_i)\hat{u}_i \quad \text{s.t.} \quad (4.13), \quad i = 1, \cdots, l.$$

This shows that Step 2 of the Piecewise strategy can be accomplished by linear programming. Rosen actually solves the dual of (4.14). His Theorems 1 and 2 concern Step 3 (cf. the discussion following (3.4) in §3.1).

It is interesting to note that if we had started with the dual of (4.9)—a block-diagonal linear program with coupling constraints—we would obtain precisely the same procedure as the one just described by dualizing with respect to the coupling constraints only [Geoffrion 69] and then invoking the Piecewise strategy. In this way [Rosen 64] could also be used to illustrate the pattern DUALIZATION/PIECEWISE.

4.3   Dantzig-Wolfe Decomposition

Dantzig-Wolfe Decomposition is archetypical of the pattern INNER LINEARIZATION/RESTRICTION. Mechanized pricing plays a prominent role. We shall illustrate this pattern first with the algorithm of [Dantzig and Wolfe 60] for a purely linear program, then with the algorithm of [Dantzig 63a, Ch. 24] for a nonlinear program, and finally with a variation of the latter in which not all nonlinear functions need be inner-linearized.

It is interesting to note that Dantzig-Wolfe Decomposition can also be viewed as an instance of the pattern DUALIZATION, OUTER LINEARIZATION/RELAXATION. In the context of (4.15), for example, one would dualize with respect to the constraints $\bar{A}x \leqq \bar{b}$, outer-linearize the resulting minimand in the obvious way, and then apply Relaxation.

[Dantzig and Wolfe 60]. The well-known Dantzig-Wolfe decomposition approach for linear programs will be explained in terms of the linear program

$$(4.15) \qquad\qquad \text{Maximize}_{x \geqq 0} \quad c^t x \quad \text{s.t.} \quad Ax \leqq b, \quad \bar{A}x \leqq \bar{b},$$

where we have arbitrarily divided the constraints into two groups. With the definition

$$(4.16) \qquad\qquad X \triangleq \{x \geqq 0 : Ax \leqq b\},$$

we may write (4.15) as

$$(4.17) \qquad\qquad \text{Maximize}_{x \in X} \quad c^t x \quad \text{s.t.} \quad \bar{A}x \leqq \bar{b}.$$

Since X is a convex polytope, we know (Theorem 3) that it admits an exact inner linearization using only a finite number of points. Invoking this representation for X, we obtain a new master linear program with a vast number of variables to which Restriction can be applied in the form of the Simplex Method. It turns out that the pricing operation (cf. §3.2) can be accomplished by solving a linear subproblem whose feasible region is X. The details are as follows.

Assume that X is not empty and also, for ease of exposition only, that X is bounded. Then X can be represented in terms of its extreme points $\langle x^1, \cdots, x^p \rangle$, and (4.17) can be written as the equivalent master linear program

$$(4.18) \quad \text{Maximize}_{\alpha \geq 0} \quad c^t(\textstyle\sum_{j=1}^p \alpha_j x^j) \quad \text{s.t.} \quad \textstyle\sum_{j=1}^p \alpha_j = 1, \quad \bar{A}(\textstyle\sum_{j=1}^p \alpha_j x^j) \leqq \bar{b}.$$

The Simplex Method for this problem corresponds to Restriction with respect to the constraints $\alpha \geqq 0$.[2] To describe how the pricing operation can be mechanized, we shall use the familiar terminology of linear programming rather than the general terminology of Restriction. The optimality conditions at the general iteration are $u \geqq 0$ and

$$(4.19) \qquad u_0 + u^t \bar{A} x^j - c^t x^j \geq 0, \qquad j = 1, \cdots, p,$$

where u_0 and the vector u are the current Simplex multipliers. Condition (4.19) is equivalent to

$$[u_0 + \text{Minimum}_{1 \leq j \leq p} \ \{(u^t \bar{A} - c^t)x^j\}] \geq 0$$

or, since $\langle x^1, \cdots, x^p \rangle$ span X, to

$$(4.20) \qquad [u_0 + \text{Min}_{x \in X} \ (u^t \bar{A} - c^t)x] \geq 0.$$

The linear program in this expression is a valid replacement for the finite minimum in the previous expression because the minimum of a linear function over X occurs at an extreme point. Thus we see how to test optimality when the Simplex Method is applied to (4.18). If either $u \geqq 0$ or (4.20) fails to hold, a profitable nonbasic variable satisfying the usual criterion for the entering variable is obtained automatically: if the greatest violation occurs in $u \geqq 0$, introduce the corresponding slack variable; if in (4.20), introduce the variable α_{j_0}, where x^{j_0} is an optimal basic feasible solution of the linear program in (4.20) (the extremal function coefficient of α_{j_0} is $c^t x^{j_0}$, and the technological coefficient column is unity followed by $\bar{A} x^{j_0}$).

Thus there is no difficulty in carrying out the Simplex Method applied to (4.18). Each iteration requires solving the linear subproblem in (4.20).[3] This approach may possess an advantage over the direct application of the Simplex Method to (4.15) when the subproblem has some special structure. For example, if (4.15) is a transportation problem with additional constraints, then the subproblem becomes a pure transportation problem if $\bar{A}$ is taken to comprise the additional constraints. Another example is the case in which A is block-diagonal, for then the subproblem separates into k independent smaller linear programs. In general, one should select a grouping of the constraints (in terms of A and $\bar{A}$) that isolates a special structure, and then exploit this structure in dealing with (4.20). See [Broise, Huard and Sentenac 68], [Orchard-Hays 68, §10.4] for additional discussion based on computational experience.

[Dantzig 63a, Chapter 24]. Now consider a nonlinear version of (4.17), namely

$$(4.21) \qquad \text{Maximize}_{x \in X} \quad f(x) \quad \text{s.t.} \quad g_i(x) \leq b_i, \qquad i = 1, \cdots, m,$$

---

[2] Actually, the inequality constraints involving $\bar{A}$ are also normally considered as candidates for restriction to equality. The latter constraints can be excluded, if desired, from the candidates for restriction by giving $u \geqq 0$ priority over (4.19) in determining the entering basic variable. Such a modification is necessary, as we shall see later in this subsection, when nonlinear functions are inner-linearized.

[3] The subproblem need be solved from scratch only at the first iteration; thereafter, restarting or parametric techniques can be used to recover an optimum as u changes from iteration to iteration.

where X is a convex set, f is concave on X, and g_i is convex on X. Dantzig and Wolfe's approach [Dantzig 63a, Chapter 24] for this problem can be viewed as follows. Let f and each g_i be approximated by Inner Linearization over an arbitrarily fine base $\langle x^1, x^2, \cdots \rangle$ in X, so that (4.21) is approximated as closely as desired (in principle, at least) by the linear master problem

$$(4.22) \qquad \text{Maximize}_{\alpha \geq 0} \ \sum_j \alpha_j f(x^j) \quad \text{s.t.} \quad \sum_j \alpha_j = 1,$$
$$\sum_j \alpha_j g_i(x^j) \leq b_i, \quad i = 1, \cdots, m.$$

We say "in principle" because we do not wish to actually evaluate f and each g_i at every point in the base, or even specify the base explicitly. Hence it is natural to solve (4.22) by Restriction with the constraints $\alpha \geq 0$ as the candidates for restriction to equality (when α_j is restricted to 0, the values $f(x^j)$ and $g_i(x^j)$ are not needed). A very natural way to do this is to employ the Simplex Method with a priority convention to ensure that the restricted problems are truly optimized: slack variables corresponding to the g_i constraints must be given priority over structural variables in determining which variable is to enter a basis. Any feasible solution of (4.21) can be used to find an initial basic feasible solution, and at the general iteration the optimality criterion or pricing problem is (cf. (4.19)) $u_i \geq 0$ ($1 \leq i \leq m$) and

$$(4.23) \qquad u_0 + \sum_{i=1}^m u_i g_i(x^j) - f(x^j) \geq 0, \quad \text{all } j,$$

where $u_0, u_1, \cdots, u_m$ are the current Simplex multipliers. By the priority convention, we may assume that $u_i \geq 0$ ($1 \leq i \leq m$). Note that (4.23) is intimately related (cf. (4.20)) to the convex subproblem

$$(4.24) \qquad \text{Minimize}_{x \in X} \ \sum_{i=1}^m u_i g_i(x) - f(x).$$

If u_0 plus the optimal value of this problem is nonnegative, then (4.23) holds and an optimal solution of (4.21) is at hand ($x^* = \sum_j \hat{\alpha}_j x^j$, where $\hat{\alpha}$ is the current and optimal solution of (4.22)); otherwise, an optimal or near-optimal solution $\hat{x}$ of (4.24) can be profitably added to the current explicit base by introducing the corresponding α_j into the basis in the usual way after evaluating $f(\hat{x})$ and $g_i(\hat{x})$. In practice, termination would take place as soon as the value of the current approximation to an optimal solution of (4.21)—the quantity $f(\sum_j \hat{\alpha}_j x_j)$—approaches closely enough the following easily demonstrated upper bound for the true optimal value:

$$(4.25) \qquad \sum_{i=1}^m u_i b_i - \text{Min}_{x \in X}[\sum_{i=1}^m u_i g_i(x) - f(x)].$$

This approach is particularly attractive when the structure is such that (4.24) is relatively tractable by comparison with (4.21); for example, when X is an open set and f and g_i are differentiable, or when (4.24) is separable into several independent subproblems.

A Variant. It is interesting to observe that Inner Linearization need not be applied to *all* nonlinear functions of (4.21). [4] An advantage can sometimes be gained by inner-linearizing only a subset of the nonlinear functions, say $g_1, \cdots, g_{m_1}$ ($m_1 < m$). Then instead of (4.22) we have the concave master problem

$$(4.26) \qquad \text{Maximize}_{\alpha \geq 0} f(\sum_j \alpha_j x^j) \quad \text{s.t.} \quad \sum_j \alpha_j = 1,$$
$$\sum_j \alpha_j g_i(x^j) \leq b_i, \quad i = 1, \cdots, m_1,$$
$$g_i(\sum_j \alpha_j x^j) \leq b_i, \quad i = m_1 + 1, \cdots, m.$$

---

[4] In [Whinston 66], for example, the objective function of a block-diagonal quadratic program with coupling constraints is not inner-linearized.

Again we wish to apply Restriction with only the nonnegativity constraints $\alpha \geq 0$ as candidates for restriction to equality. The Simplex Method can no longer be adapted to this purpose, however, since (4.26) is not a linear program. Implementation requires a concave programming algorithm for solving the restricted versions of (4.26) and also a means of mechanizing the pricing operation. We need not discuss the first requirement. The second involves being able to determine the prices $\mu_j{}^s$ for all j in S, where S is the current set of indices for which α_j is restricted to value 0. This can be done as follows [Holloway 69]. Let α^s be the optimal solution to (4.26) with the additional restrictions $\alpha_j = 0$ for $j \in S$, and let $u_0{}^s, u_1{}^s, \cdots, u_m{}^s$ be the associated optimal multipliers (which must exist if a constraint qualification is satisfied). Then, assuming all functions are continuously differentiable, the price $\mu_j{}^s$ associated with $\alpha_j = 0$ is given for all $j \in S$ by

$$(4.27) \qquad \mu_j{}^s = u_0{}^s - \nabla f(x^s)x^j + \sum_{i=1}^{m_1} u_i{}^s g_i(x^j) + \sum_{i=m_1+1}^{m} u_i{}^s \nabla g_i(x^s)x^j,$$

where

$$(4.28) \qquad x^s \triangleq \sum_{j \in S} \alpha_j{}^s x^j.$$

It follows that the pricing problem can be solved by optimizing the convex $(u_i{}^s \geq 0)$ subproblem

$$(4.29) \quad \text{Minimize}_{x \in X} \quad - \nabla f(x^s)x + \sum_{i=1}^{m_1} u_i{}^s g_i(x) + \sum_{i=m_1+1}^{m} u_i{}^s \nabla g_i(x^s)x.$$

Compare with (4.24). If f were inner-linearized too, the first term of the maximand of (4.29) would be $-f(x)$.

Which of all given constraints should be incorporated into X, and which of the remainder and whether f itself should be inner-linearized, depends mainly on the availability of efficient algorithms for the resulting versions of (4.29) and (4.26) with $\alpha_j = 0$ for $j \in S$.

4.4 [Geoffrion 68b; §4]

A quite general problem with multidivisional structure is

$$(4.30) \qquad \text{Maximize}_x \quad \sum_{i=1}^{k} f_i(x_i) \quad \text{s.t.} \quad H_i(x_i) \geq 0, \quad i = 1, \cdots, k$$
$$\sum_{i=1}^{k} G_i(x_i) \geq b,$$

where f_i, h_{ij} and g_{ij} are all concave differentiable functions of the vector x_i. The subscript i can be thought of as indexing the individual divisions, which are linked together only by coupling constraints. The approach of [Geoffrion 68b; §4] is an application of the pattern PROJECTION/FEASIBLE DIRECTIONS. The optimization of (4.30) is carried out largely at the divisional level subject to central coordination.

First (4.30) is projected onto the space of its coupling constraints. This requires introducing the vectors $y_1, \cdots, y_k$:

$$(4.31) \qquad \text{Maximize}_{x,y} \quad \sum_{i=1}^{k} f_i(x_i) \quad \text{s.t.}$$
$$H_i(x_i) \geq 0, i = 1, \cdots, k; \ G_i(x_i) \geq y_i, i = 1, \cdots, k; \ \sum_{i=1}^{k} y_i \geq b.$$

In effect, this changes the given problem from one with coupling constraints to one with coupling variables, since (4.31) separates into k separate problems if y is held fixed temporarily. One may interpret y_i as a vector of resources and tasks assigned to the ith division. Projection of this problem onto y yields the master problem

$$(4.32) \qquad \text{Maximize}_y \quad \sum_{i=1}^{k} v_i(y_i) \quad \text{s.t.} \quad \sum_{i=1}^{k} y_i \geq b,$$

where v_i is defined as the supremal value of the parameterized divisional problem

$$(4.33) \qquad \text{Maximize}_{x_i} \quad f_i(x_i) \quad \text{s.t.} \quad (H_i(x_i) \geq 0, \qquad G_i(x_i) \geq y_i.$$

Now we wish to apply the Feasible Directions strategy to (4.32). The idea of this strategy, it will be recalled, is to generate an improving sequence of feasible points, with each new point determined from the previous one by selecting an improving feasible direction and then maximizing along a line emanating in this direction. The latter maximization is only one-dimensional, and can easily be essentially decentralized to the divisional level. The chief difficulty with this strategy concerns how to find a good improving feasible direction, for the maximand $\sum_{i=1}^{k} v_i(y_i)$ is not everywhere differentiable and is available only implicitly in terms of the divisional problems (4.33). It can nevertheless be shown [ibid., §4.2], using the theory of subgradients for concave functions and the optimality conditions associated with (4.33), that the following explicit linear program yields an improving feasible direction z^0 for (4.32) at a feasible point y^0; moreover, z^0 is *best* among all feasible directions in that it maximizes the initial rate of improvement of $\sum_{i=1}^{k} v_i(y_i)$:

$$\text{Maximize}_{w,z} \quad \sum_{i=1}^{k} \nabla f_i^0 w_i \quad \text{s.t.}$$

$$\nabla g_{ij}^0 w_i - z_{ij} \geq 0, \quad i = 1, \cdots, k$$
$$j \text{ such that } g_{ij}^0 = y_{ij}^0$$

$$\nabla h_{ij}^0 w_i \geq 0, \quad i = 1, \cdots, k$$

(4.34)
$$j \text{ such that } h_{ij}^0 = 0$$

$$\sum_{i=1}^{k} z_{ij} \geq 0, \quad j \text{ such that } \sum_{i=1}^{k} y_{ij}^0 = b_j$$

$$-1 \leq z_{ij} \leq 1, \quad \text{all } i \text{ and } j.$$

Here ∇g_{ij}^0 refers to a row vector that is the gradient of g_{ij} evaluated at an optimal solution of (4.33) with $y_i = y_i^0$, and the other superscripted quantities have similar definitions. The vector w_i has the same dimension as x_i. This subproblem enables the Feasible Directions strategy for (4.32) to be carried out.

4.5 [*Takahashi* 64]

Consider

(4.35) $\text{Maximize}_x \quad f(x) \quad \text{s.t.} \quad H(x) = 0, \quad G(x) = 0,$

where f is concave and all constraints are linear. Suppose that the G constraints are *complicating* in the sense that the problem would be much easier if they were not present. For instance, the complicating constraints may be the coupling constraints of a structure similar to the one in the previous subsection, or they may spoil what would otherwise be a special structure for which efficient solution methods would be available. The pattern of the "local" approach of [Takahashi 64] for this problem is DUALIZATION/FEASIBLE DIRECTIONS.

The dual of (4.35) with respect to the complicating constraints only yields (see, e.g., [Rockafellar 68] or (Geoffrion 69]) the following problem in the space of the dual variables λ (a vector whose dimension matches G):

(4.36) $\text{Minimize}_\lambda \quad v(\lambda),$

where $v(\lambda)$ is defined as the supremal value of the parameterized problem

(4.37) $\text{Maximize}_x \quad f(x) + \lambda' G(x) \quad \text{s.t.} \quad H(x) = 0.$

Note that (4.37) is of the same form as (4.35) except the complicating constraints are now part of the criterion function.

To apply the Feasible Directions strategy to (4.36), we must be able to identify an improving feasible direction. *Any* direction is feasible, of course, since λ is unconstrained. When f is strictly concave, it can be shown that v is differentiable. Its gradient at a point λ^0 is simply $G(x^0)$, where x^0 is the optimal solution of (4.37) with $\lambda = \lambda^0$. Hence the Feasible Directions strategy can be carried out for (4.36) using the negative of the gradient of v as the improving feasible direction. Actually, Takahashi proposes a short-step method rather than requiring a one-dimensional minimization to be performed in order to determine step size. The procedure may be summarized as follows.

1. Choose a starting point λ^0.

2. Solve (4.37) with $\lambda = \lambda^0$ for its optimal solution x^0. If $G(x^0) = 0$, then x^0 is optimal in (4.35); otherwise, go on to Step 3.

3. Let $\lambda' = \lambda^0 - \zeta G(x^0)$, where ζ is a small positive constant, and return to Step 2 with λ' in place of λ^0.

5. Conclusion

We have attempted to develop a framework of unifying concepts that comprehends much of the literature on large-scale mathematical programmning. If we have been successful, the nonspecialist should have an overview of the field that facilitates further study, and the advanced reader should feel that he has a deeper understanding of previously familiar algorithms and that he perceives new commonalities among approaches that heretofore seemed to be related only vaguely if at all.

In addition, we hope that the framework will suggest a variety of worthwhile topics for investigation. The problem manipulations and solution strategies discussed here all invite further study, and others should be added to the fold so that additional algorithms can be encompassed. The algorithms falling within the purview of each particular manipulation/strategy pattern (cf. Table 2) should be studied carefully in relation to one another, with the aim of learning how "best" to use the tactical options of the pattern and organize the computations for various classes of problems.

The relationships between ostensibly different patterns also warrant further study. We mentioned in §3.3 that Restriction (Relaxation) is essentially equivalent to Dualization followed by Relaxation (Restriction), and other equivalences were briefly noted in §4.2 and §4.3. Many others exist; for example, it has often been observed that Dantzig-Wolfe and Benders Decomposition are dual to one another in an appropriate sense. The results of [Zoutendijk 60; Secs. 9.4, 10.3, 11.4] are in this spirit, even if they do not specifically involve algorithms for large-scale programming. Knowledge of such relations reduces the number of essentially different patterns to be considered, and enables meaningful comparisons among the remainder.

Investigations along these lines should help civilize the jungle of extant algorithms and pave the way for truly significant computational studies.

BIBLIOGRAPHY[5]

ABADIE, J. and SAKAROVITCH, M., "Two Methods of Decomposition for Linear Programs," presented at the International Symposium on Mathematical Programming, Princeton, New Jersey, August 14–18, 1967.

—— and WILLIAMS, A. C., "Dual and Parametric Methods in Decomposition," in R. L. Graves and P. Wolfe (eds.), *Recent Advances in Mathematical Programming*, McGraw-Hill Book Company, New York, 1963.

---

[5] Items marked * are not ordinarily considered part of the large-scale mathematical programming literature.

AOKI, M., "Planning Procedure Under Increasing Returns," Harvard Economic Research Project, Harvard University, Cambridge, 1968.

APPELGREN, L. H., "A Column-Generation Algorithm for a Ship Scheduling Problem," *Transportation Science*, Vol. 3, No. 1 (February 1969), pp. 53–68.

BAKES, M. D., "Solution of Special Linear Programming Problems with Additional Constraints," *Operational Research Quarterly*, Vol. 17, No. 4 (December 1966), pp. 425–445.

BALAS, E., "Solution of Large Scale Transportation Problems Through Aggregation," *Operations Research*, Vol. 13, No. 1 (January-February 1965), pp. 82–93.

——, "An Infeasibility-Pricing Decomposition Method for Linear Programs," *Operations Research*, Vol. 14, No. 5 (September–October 1966), pp. 847–873.

BALINSKI, M. L., "On Some Decomposition Approaches in Linear Programming," in *Recent Mathematical Advances in Operations Research*, Engineering Summer Conferences, University of Michigan, 1964.

—— AND WOLFE, P., "On Benders Decomposition and a Plant Location Problem," Working Paper ARO-27, Mathematica, Princeton, New Jersey, December 1963.

BAUMOL, W. J. AND FABIAN, T., "Decomposition, Pricing for Decentralization and External Economies," *Management Science*, Vol. 11, No. 1 (September 1964), pp. 1–32.

BEALE, E. M. L., "The Simplex Method Using Pseudo-basic Variables for Structured Linear Programs," in R. L. Graves and P. Wolfe (eds.), *Recent Advances in Mathematical Programming*, McGraw-Hill Book Company, New York, 1963.

——, "Decomposition and Partitioning Methods for Nonlinear Programming," in J. Abadie (ed.), *Non-Linear Programming*, North-Holland Publishing Company, Amsterdam, 1967, pp. 197–205.

BELL, E. J., "Primal-Dual Decomposition Programming," *Preprints of the Proceedings of the Fourth International Conference on Operational Research*, International Federation of Operational Research Societies, Boston, 1966.

BELLMAN, R., "On the Computational Solution of Linear Programming Problems Involving Almost-Block-Diagonal Matrices," *Management Science*, Vol. 3, No. 4 (July 1957), pp. 403–406.

BENDERS, J. F., "Partitioning Procedures for Solving Mixed-Variables Programming Problems," *Numerische Mathematik*, Vol. 4 (1962), pp. 238–252.

BENNETT, J. M., "An Approach to Some Structured Linear Programming Problems," *Operations Research*, Vol. 14, No. 4 (July–August 1966), pp. 636–645.

—— AND D. R. GREEN, "A Method for Solving Partially Linked Linear Programming Problems," Basser Computing Department Technical Report No. 38, University of Sydney, February 1966.

*BERGE, C., *Topological Spaces*, Macmillan Company, New York, 1963.

*—— AND GHOUILA-HOURI, A., *Programming, Games and Transportation Networks*, John Wiley and Sons, New York, 1965.

BESSIERE, F., AND SAUTTER, E. A., "Optimization and Suboptimization: The Method of Extended Models in the Nonlinear Case," *Management Science*, Vol. 15, No. 1 (September 1968), pp. 1–11.

BRADLEY, S. P., "Solution Techniques for the Traffic Assignment Problem," Operations Research Center Report 65-35, University of California, Berkeley, June 1965.

——, "Decomposition Programming and Economic Planning," Operations Research Center Report 67–20, University of California, Berkeley, June 1967.

——, "Nonlinear Programming Via the Conjugate Dual," IBM Report, Federal Systems Center, Gaithersburg, Maryland, June 1968.

BROISE, P., HUARD, P. AND SENTENAC, J., "Décomposition des programmes mathématiques," *Monographies de Recherche Opérationnelle*, No. 6, Dunod, Paris, 1968.

BUCHET, J. DE, "Expériences et Statistiques Sur la Résolution des Programmes Linéaires de Grandes Dimensions," reprinted in D. B. Hertz and J. Melese (eds.), *Proceedings of the Fourth International Conference on Operational Research*, John Wiley and Sons, New York, 1966.

BUZBY, B. R., STONE, B. J., AND TAYLOR, R. L., "Computational Experience with a Nonlinear Distribution Program," privately communicated to M. L. Balinski, January 1965. See M. L. Balinski, "Integer Programming: Methods, Uses, Computation," *Management Science*, Vol. 12, No. 3 (November 1965), p. 308.

CHANDY, K. M., "Parametric Primal Partitioning," Operations Research Center, Massachusetts Institute of Technology, Cambridge, 1968.

CHARNES, A., CLOWER, R. W., AND KORTANEK, K. O., "Effective Control Through Coherent Decentralization with Preemptive Goals," *Econometrica*, Vol. 35, No. 2 (April 1967), pp. 294–320.

—— AND COOPER, W. W., "Generalizations of the Warehousing Model," *Operational Research Quarterly*, Vol. 6, No. 4 (December 1955), pp. 131–172.

——, —— AND KORTANEK, K. O., "On the Theory of Semi-Infinite Programming and a Generalization of the Kuhn-Tucker Saddle Point Theorem for Arbitrary Convex Functions," *Naval Research Logistics Quarterly*, Vol. 16, No. 1 (March 1969), pp. 41–51.

—— AND KORTANEK, K. O., "On the Status of Separability and Non-Separability in Decentralization Theory," *Management Science*, Vol. 15, No. 2 (October 1968), pp. 12–14.

—— AND LEMKE, C., "Minimization of Nonlinear Separable Convex Functionals," *Naval Research Logistics Quarterly*, Vol. 1 (1954), pp. 301–312.

*CHENEY, E. W. AND GOLDSTEIN, A. A., "Newton's Method for Convex Programming and Tchebycheff Approximation," *Numerische Mathematik*, Vol. 1 (1959), pp. 253–268.

COBB, R. H. AND CORD, J., "Decomposition Approaches for Solving Linked Programs," presented at the International Symposium on Mathematical Programming, Princeton, New Jersey, August 14–18, 1967.

DANTZIG, G. B., "Upper Bounds, Secondary Constraints, and Block Triangularity in Linear Programming," *Econometrica*, Vol. 23, No. 2 (April 1955a), pp. 174–183.

——, "Optimal Solution of a Dynamic Leontief Model with Substitution," *Econometrica*, Vol. 23, No. 3 (July 1955b), pp. 295–302.

——, "On the Status of Multistage Linear Programming Problems," *Management Science*, Vol. 6, No. 1 (October 1959), pp. 53–72.

——, "A Machine-Job Scheduling Model," *Management Science*, Vol. 6, No. 2 (January 1960), pp. 191–196.

——, *Linear Programming and Extensions*, Princeton University Press, Princeton, New Jersey, 1963a.

——, "Compact Basis Triangularization for the Simplex Method," in R. Graves and P. Wolfe (eds.), *Recent Advances in Mathematical Programming*, McGraw-Hill Book Company, New York, 1963b.

——, "Linear Control Processes and Mathematical Programming," *SIAM Journal on Control*, Vol. 4, No. 1 (February 1966), pp. 56–60.

——, "Large-Scale Linear Programming," in G. B. Dantzig and A. F. Veinott, Jr. (eds.), *Mathematics of the Decision Sciences*, Part 1, American Mathematical Society, Providence, 1968.

——, BLATTNER, W., AND RAO, M. R., "Finding a Cycle in a Graph with Minimum Cost to Time Ratio with Applications to a Ship Routing Problem," *Theory of Graphs International Symposium*, Dunod, Paris, 1967, pp. 77–83.

——, FULKERSON, D. R. AND JOHNSON, S., "Solution of a Large-Scale Traveling Salesman Problem," *Operations Research*, Vol. 2, No. 4 (November 1954), pp. 393–410.

——, ——, AND ——, "On a Linear Programming Combinatorial Approach to the Traveling-Salesman Problem," *Operations Research*, Vol. 7, No. 1 (January–February 1959), pp. 58–66.

——, HARVEY, R. P., AND McKNIGHT, R. D., "Updating the Product Form of the Inverse for the Revised Simplex Method," Operations Research Center Report 64-33, University of California, Berkeley, December 1964.

—— AND JOHNSON, D. L., "Maximum Payload Per Unit Time Delivered Through an Air Network," *Operations Research*, Vol. 12, No. 2 (March–April 1964), pp. 230–236.

—— AND MADANSKY, A., "On the Solution of Two-Stage Linear Programs Under Uncertainty," in *Proceedings of the Fourth Symposium on Mathematical Statistics and Probability*, Vol. I, University of California Press, Berkeley, 1961, pp. 165–176.

—— AND ORCHARD-HAYS, W., "The Product Form of the Inverse in the Simplex Method," *Mathematical Tables and Aids to Computation*, Vol. 8 (1954), pp. 64–67.

—— AND VAN SLYKE, R. M., "Generalized Upper Bounded Techniques for Linear Programming," *Journal of Computer and System Sciences*, Vol. 1, No. 3 (October 1967), pp. 213–226.

—— AND WOLFE, P., "Decomposition Principle for Linear Programs," *Operations Research*, Vol. 8, No. 1 (January–February 1960), pp. 101–111. See also *Econometrica*, Vol. 29, No. 4 (October), pp. 767–778.

DZIELINSKI, B. P. AND GOMORY, R. E., "Optimal Programming of Lot Sizes, Inventory and Labor Allocations," *Management Science*, Vol. 11, No. 9 (July 1965), pp. 874–890.

*EAVES, B. C., AND ZANGWILL, W. I., "Generalized Cutting Plane Algorithms," Working Paper No. 274, Center for Research in Management Science, University of California, Berkeley, July 1969.

EL AGIZY, M., "Two-Stage Programming Under Uncertainty with Discrete Distribution Function," *Operations Research*, Vol. 15, No. 1 (January–February 1967), pp. 55–70.

ELMAGHRABY, S. E., "A Loading Problem in Process Type Production," *Operations Research*, Vol. 16, No. 5 (September–October 1968), pp. 902–914.

*FALK, J., "An Algorithm for Separable Convex Programming under Linear Equality Constraints," Research Analysis Corporation Technical Paper 148, McLean, Virginia, March 1965.

*——, "Lagrange Multipliers and Nonlinear Programming," *Journal of Mathematical Analysis and Application*, Vol. 19, No. 1 (1967), pp. 141–159.

*FIACCO, A. V. AND McCORMICK, G. P., *Nonlinear Programming*, John Wiley and Sons, Inc., New York, 1968.

FORD, L. R. AND FULKERSON, D. R., "A Suggested Computation for Maximal Multi-Commodity Network Flows," *Management Science*, Vol. 5, No. 1 (October 1958), pp. 97–101.

FOX, B., "Finding Minimal Cost-Time Ratio Circuits," *Operations Research*, Vol. 17, No. 3 (May–June 1969), pp. 546–551.

——, *Stability of the Dual Cutting-Plane Algorithm for Concave Programming*, The RAND Corporation, RM-6147-PR, February, 1970.

GASS, S. I., "The Dualplex Method for Large-Scale Linear Programs," Operations Research Center Report 66–15, University of California, Berkeley, June 1966.

GEOFFRION, A. M., "A Markovian Procedure for Concave Programming with Some Linear Constraints," reprinted in D. B. Hertz and J. Melese (eds.), *Proceedings of the Fourth International Conference on Operational Research*, John Wiley and Sons, New York, 1966.

——, "Reducing Concave Programs with Some Linear Constraints," *SIAM Journal on Applied Mathematics*, Vol. 15, No. 3 (May 1967), pp. 653–664.

——, "Relaxation and the Dual Method in Mathematical Programming," Western Management Science Institute Working Paper No. 135, University of California, Los Angeles, March 1968a.

——, "Primal Resource-Directive Approaches for Optimizing Nonlinear Decomposable Systems," the RAND Corporation, RM-5829-PR, December 1968b; to appear in *Operations Research*.

——, "Duality in Nonlinear Programming: A Simplified Applications-Oriented Development," Western Management Science Institute Working Paper No. 150, University of California, Los Angeles, August 1969; to appear in *SIAM Review*.

——, "Generalized Benders Decomposition," *Proceedings of the Symposium on Nonlinear Programming*, Mathematics Research Center, University of Wisconsin, Madison, May 4–6, 1970.

GILMORE, P. C. AND GOMORY, R. E., "A Linear Programming Approach to the Cutting-Stock Problem," *Operations Research*, Vol. 9, No. 6 (November–December 1961), pp. 849–859.

—— AND ——, "A Linear Programming Approach to the Cutting-Stock Problem—Part II," *Operations Research*, Vol. 11, No. 6 (November–December 1963), pp. 863–888.

—— AND ——, "Multistage Cutting Stock Problems of Two and More Dimensions," *Operations Research*, Vol. 13, No. 1 (January–February 1965), pp. 94–120.

GLASSEY, C. R., "An Algorithm for a Machine Loading Problem," Operations Research Center Working Paper 30, University of California, Berkeley, September 1966.

——, "Dynamic Linear Programs for Production Scheduling," Operations Research Center Report 68–2, University of California, Berkeley, January 1968.

*GOLDMAN, A. J., "Resolution and Separation Theorems for Polyhedral Convex Sets," in

H. W. Kuhn and A. W. Tucker (eds.), *Linear Inequalities and Related Systems*, Princeton University Press, Princeton, New Jersey, 1956.

GOLSHTEIN, E. G. "A General Approach to the Linear Programming of Block Structures," *Soviet Physics—Doklady*, Vol. 11, No. 2 (August 1966), pp. 100–103.

*GOMORY, R. E., 1958 "An Algorithm for Integer Solutions to Linear Programs," reprinted in R. L. Graves and P. Wolfe (eds.), *Recent Advances in Mathematical Programming*, McGraw-Hill Book Company, New York, 1936.

——, "Large and Nonconvex Problems in Linear Programming," in *Proceedings of Symposia in Applied Mathematics*, Vol. XV (1963), American Mathematical Society, pp. 125–139.

—— AND HU, T. C., "An Application of Generalized Linear Programming to Network Flows," *Journal of the Society for Industrial and Applied Mathematics*, Vol. 10, No. 2 (June 1962), pp. 260–283.

—— AND ——, "Synthesis of a Communication Network," *Journal of the Society for Industrial and Applied Mathematics*, Vol. 12, No. 2 (June 1964), pp. 348–369.

GOULD, S., "A Method of Dealing with Certain Non-Linear Allocation Problems Using the Transportation Technique," *Operational Research Quarterly*, Vol. 10, No. 3 (September 1959), pp. 138–171.

GRAVES, G. W., HATFIELD, G. B., AND WHINSTON, A., "Water Pollution Control Using By-Pass Piping," *Water Resources Research*, Vol. 5, No. 1 (February 1969).

GRIGORIADIS, M. D., "A Dual Generalized Upper Bounding Technique," IBM New York Scientific Center Report No. 320-2973, New York, June 1969.

—— AND RITTER, K., "A Decomposition Method for Structured Linear and Non-Linear Programs," *J. Computer and System Sciences*, Vol. 3, No. 4 (November 1969), pp. 335–360.

—— AND WALKER, W. F., "A Treatment of Transportation Problems by Primal Partition Programming," *Management Science*, Vol. 14, No. 9 (May, 1968), pp. 565–599.

*GRINOLD, R. C., "Continuous Programming," Operations Research Center Report 68-14, University of California, Berkeley, June 1968.

——, "Steepest Ascent for Large-Scale Linear Programs," Center for Research in Management Science Working Paper No. 282, University of California, Berkeley, September 1969.

*HARTLEY, H. O. AND HOCKING, R. R., "Convex Programming by Tangential Approximation," *Management Science*, Vol. 9, No. 4 (July 1963), pp. 600–612.

HASS, J. E., "Transfer Pricing in a Decentralized Firm," *Management Science*, Vol. 14, No. 6 (February 1968), pp. B-310-B-331.

HEESTERMAN, A. R. G. AND SANDEE, J. "Special Simplex Algorithm for Linked Problems," *Management Science*, Vol. 11, No. 3 (January 1965), pp. 420–428.

HELD, M., AND KARP, R. M., "The Traveling-Salesman Problem and Minimum Spanning Trees," IBM Systems Research Institute, August 1969.

HOLLOWAY, C. A., "A Mathematical Programming Approach to Decision Processes for Complex Operational Systems, With an Application to the Aggregate Planning Problem," Ph.D. dissertation, Graduate School of Business Administration, University of California, Los Angeles, 1969.

HOPKINS, D. S. P., "Sufficient Conditions for Optimality in Infinite Horizon Linear Economic Models," Operations Research House Technical Report 69-3, Stanford University, March 1969.

*JOKSCH, H. C., "Programming with Fractional Linear Objective Functions," *Naval Research Logistics Quarterly*, Vol. 11, Nos. 2 & 3 (June–September 1964), pp. 197–204.

KAUL, R. N., "Comment on 'Generalized Upper Bounded Techniques in Linear Programming'," Operations Research Center Report 65-21 (August), University of California, Berkeley, 1965a.

——, "An Extension of Generalized Upper Bounded Techniques for Linear Programming," Operations Research Center Report 65-27 (August), University of California, Berkeley, 1965b.

*KELLEY, J. E., JR., "The Cutting-Plane Method for Solving Convex Programs, *Journal of the Society for Industrial and Applied Mathematics*, Vol. 8, No. 4 (December 1960), pp. 703–712.

*KOHLER, D. A., "Projections of Convex Polyhedral Sets," Operations Research Center Report 67-29, University of California, Berkeley, August 1967.

KORNAI, J. AND LIPTAK, T., "Two-Level Planning," *Econometrica*, Vol. 33, No. 1 (January 1965), pp. 141–169.

KRONSJÖ, T. O. M., "Centralization and Decentralization of Decision Making," *Revue d'Informatique et de Recherche Operationelle*, Vol. 2, No. 10 (1968a), pp. 73–114.

——, "Optimal Coordination of a Large Convex Economic System," Discussion Paper Series RC/A, No. 9, University of Birmingham, Faculty of Commerce and Social Sciences, Birmingham, Great Britain, January 1968b.

KUNZI, H. P., AND TAN, S. T., *Lineare Optimierung Grosser Systeme*, Springer-Verlag, Berlin, 1966.

LASDON, L. S., "A Multi-Level Technique for Optimization," Systems Research Center Report 50-64-19, Case Institute of Technlology, Cleveland, 1964.

——, "Duality and Decomposition in Mathematical Programming," *IEEE Transactions on System Science and Cybernetics*, Vol. SSC-4, No. 2 (July 1968), pp. 86–100.

——, *Optimization Theory for Large Systems*, The Macmillan Company, New York, 1970.

——, AND MACKEY, J. E., "An Efficient Algorithm for Multi-Item Scheduling," Technical Memorandum No. 108, Department of Operations Research, Case Western Reserve University, Cleveland, May 1968.

——, AND SCHOEFFLER, J. D., "Decentralized Plant Control," *I.S.A.* Transactions, Vol. 5 (April 1966), pp. 175–183.

MADGE, D. N., "Decomposition of Mine Scheduling Problems Involving Mining Sequence Restrictions," *C.O.R.S. Journal*, Vol. 3, No. 3 (November 1965), pp. 161–165.

MALINVAUD, E., "Decentralized Procedures for Planning," in E. Malinvaud and M. O. L. Bacharach (eds.), *Activity Analysis in the Theory of Economic Growth*, Macmillan Company, New York, 1967.

MANNE, A. S. AND MARKOWITZ, H. M. (eds.), *Studies in Process Analysis*, John Wiley and Sons, New York, 1963.

MARCEAU, I. W., LERMIT, R. J., McMILLAN, J. C., YAMAMOTO, J. AND YARDLEY, S. S., "Linear Programming on Illiac IV," Illiac IV Document No. 225 of Computer Science, University of Illinois at Urbana-Champaign, September 1969.

MARKOWITZ, H., "The Elimination Form of the Inverse and Its Application to Linear Programming," *Management Science*, Vol. 3, No. 3 (April 1957), pp. 255–269.

METZ, C. K. C., HOWARD, R. N., AND WILLIAMSON, J. M., "Applying Benders Partitioning Method to a Non-Convex Programming Problem," reprinted in D. B. Hertz and J. Melese (eds.), *Proceedings of the Fourth International Conference on Operational Research*, John Wiley and Sons, New York, 1966.

MIDLER, J. L., AND WOLLMER, R. D., "Stochastic Programming Models for Scheduling Airlift Operations," *Naval Research Logistics Quarterly*, Vol. 16, No. 3 (September 1969), pp. 315–330.

*MILLER, C. E., "The Simplex Method for Local Separable Programming," in R. L. Graves and P. Wolfe (eds.), *Recent Advances in Mathematical Programming*, McGraw-Hill Book Company, New York, 1963.

NEMHAUSER, G. L., "Decomposition of Linear Programs by Dynamic Programming," *Naval Research Logistics Quarterly*, Vol. 11, Nos. 2 & 3 (June–September 1964), pp. 191–196.

ORCHARD-HAYS, W., *Advanced Linear Programming Computing Techniques*, McGraw-Hill Book Company, New York, 1968.

*ORDEN, A. AND NALBANDIAN, V., "A Bidirectional Simplex Algorithm," *Journal of the Association for Computing Machinery*, Vol. 15, No. 2 (April 1968), pp. 221–235.

*PARIKH, S. C., "Generalized Stochastic Programs with Deterministic Recourse," Operations Research Center Report 67-27, University of California, Berkeley, July 1967.

—— AND SHEPHARD, R. W., "Linear Dynamic Decomposition Programming Approach to Long-Range Optimization of Northern California Water Resources System, Part I: Deterministic Hydrology," Operations Research Center Report 67-30, University of California, Berkeley, August 1967.

PEARSON, J. D., "Decomposition, Coordination, and Multi-level Systems," *IEEE Trans. on Systems Science and Cybernetics*, Vol. SSC-2, No. 1 (August 1966), pp. 36–40.

RAO, M. R., "Multi-Commodity Warehousing Models with Cash-Liquidity Constraints—A Decomposition Approach," Management Sciences Research Report No. 141, Graduate School of Industrial Administration, Carnegie-Mellon University, Pittsburgh, September 1968.

—— AND ZIONTS, S., "Allocation of Transportation Units to Alternative Trips—A Column Generation Scheme With Out-of-Kilter Subproblems," *Operations Research*, Vol. 16, No. 1 (January–February 1968), pp. 52–63.

RECH, P., "Decomposition of Interconnected Leontief Systems by Square Block Triangularization," *Pre-prints of the Proceedings of the Fourth International Conference on Operational Research*, International Federation of Operational Research Societies, Boston, 1966.

*RITTER, K., "A Parametric Method for Solving Certain Nonconcave Maximization Problems," *Journal of Computer and System Sciences*, Vol. 1, No. 1 (April 1967a), pp. 44–54.

——, "A Decomposition Method for Structured Quadratic Programming Problems," *Journal of Computer and System Sciences*, Vol. 1, No. 3 (October 1967b), pp. 241–260.

——, "A Decomposition Method for Linear Programming Problems with Coupling Constraints and Variables," Mathematics Research Center Technical Summary Report No. 739, University of Wisconsin, Madison, April 1967c.

ROBERS, P. D., AND BEN-ISRAEL, A., "A Decomposition Method for Interval Linear Programming," *Management Science*, Vol. 16, No. 5 (January 1970), pp. 374–387.

*——, AND ——, "On the Theory and Applications of Interval Linear Programming," Research Analysis Corporation, Tech. Paper RAC-TP-379, McLean, Virginia, October 1969.

ROBERT, J. E., "A Method of Solving a Particular Type of Very Large Linear Programming Problem," *C.O.R.S. Journal*, Vol. 1, No. 1 (December 1963), pp. 50–59.

*ROCKAFELLAR, R. T., "Duality in Nonlinear Programming," in G. B. Dantzig and A. F. Veinott, Jr. (eds.), *Mathematics of the Decision Sciences*, Part 1, American Mathematical Society, Providence, 1968.

ROSEN, J. B., "Convex Partition Programming," in R. L. Graves and P. Wolfe (eds.), *Recent Advances in Mathematical Programming*, McGraw-Hill Book Company, New York, 1963.

——, "Primal Partition Programming for Block Diagonal Matrices," *Numerische Mathematik*, Vol. 6 (1964), pp. 250–260.

——, "Optimal Control and Convex Programming," in J. Abadie (ed.), *Nonlinear Programming*, North-Holland Publishing Company, Amsterdam, 1967.

—— AND ORNEA, J. C., "Solution of Nonlinear Programming Problems by Partitioning," *Management Science*, Vol. 10, No. 1 (October 1963) pp. 160–173.

RUTENBERG, D. P., "Generalized Networks, Generalized Upper Bounding and Decomposition of the Convex Simplex Method," *Management Science*, Vol. 16, No. 5 (January 1970), pp. 388–401.

SAIGAL, R., "Compact Basis Triangularization for the Block Angular Structures," Operations Research Center Report 66-1, University of California, Berkeley, January 1966a.

——, "Block-Triangularization of Multi-Stage Linear Programs," Operations Research Center Report 66-9, University of California, Berkeley, March 1966b.

SAKAROVITCH, M. AND SAIGAL, R., "An Extension of Generalized Upper Bounding Techniques for Structured LP Problems," *SIAM Journal on Applied Mathematics*, Vol. 15, No. 4 (July 1967), pp. 906–914.

SILVERMAN, G. J., "Primal Decomposition of Mathematical Programs by Resource Allocation," Technical Memorandum No. 116, Operations Research Department, Case Western Reserve University, Cleveland, August 1968.

*STONE, J. J., "The Cross-Section Method: An Algorithm for Linear Programming," The RAND Corporation, P-1490, September 1958.

TAKAHASHI, I., "Variable Separation Principle for Mathematical Programming," *Journal of the Operations Research Society of Japan*, Vol. 6, No. 1 (February 1964), pp. 82–105.

TCHENG, T., "Scheduling of a Large Forestry-Cutting Problem by Linear Programming Decomposition," unpublished Ph.D. thesis, University of Iowa, August, 1966.

THOMPSON, G. L., TONGE, F. M., AND ZIONTS, S., "Techniques for Removing Nonbinding Constraints and Extraneous Variables from Linear Programming Problems," *Management Science*, Vol. 12, No. 7 (March 1966), pp. 588–608.

TOMLIN, J. A., "Minimum-Cost Multicommodity Network Flows," *Operations Research*, Vol. 14, No. 1 (January–February 1966), pp. 45–51.

*TOPKIS, D. M., "Cutting Plane Methods Without Nested Constraint Sets," Operations Research Center Report 69-14, University of California, Berkeley, June 1969.

*UZAWA, H., "Iterative Methods for Concave Programming," Chap. 10 in K. Arrow, L. Hurwicz and H. Uzawa (eds.), *Studies in Linear and Nonlinear Programming*, Stanford University Press, Stanford, California, 1958.

VAN SLYKE, R. M., "Mathematical Programming and Optimal Control," Operations Research Center Report 68-21, University of California, Berkeley, July 1968.

—— AND WETS, R. J. B., "L-Shaped Linear Programs with Applications to Optimal Control and Stochastic Programming," *SIAM J. Appl. Math.* Vol. 17, No. 4 (July 1969), pp. 638–663.

VARAIYA, P. "Decomposition of Large-Scale Systems," *SIAM Journal on Control*, Vol. 4, No. 1 (February 1966), pp. 173–178.

*VEINOTT, A. F., JR., "The Supporting Hyperplane Method for Unimodal Programming," *Operations Research*, Vol. 15, No. 1 (January–February 1967), pp. 147–152.

WAGNER, H. M., "A Linear Programming Solution to Dynamic Leontief Type Models," *Management Science*, Vol. 3, No. 3 (April 1957), pp. 234–254.

WEBBER, D. W. AND WHITE, W. W., "An Algorithm for Solving Large Structured Linear Programming Problems," IBM New York Scientific Center Report No. 320-2946, New York, April 1968.

WEIL, R. L., JR. AND KETTLER, P. C., "Transforming Matrices to Use the Decomposition Algorithm for Linear Programs," Report 6801, Department of Economics and Graduate School of Business, University of Chicago, January 1968.

WEITZMAN, M., "Iterative Multi-Level Planning with Production Targets," Cowles Foundation Discussion Paper No. 239, Yale University, New Haven, November 1967. To appear in *Econometrica*.

WHINSTON, A., "A Dual Decomposition Algorithm for Quadratic Programming," *Cahiers du Centre d'Etudes de Recherche Operationnelle*, Vol. 6, No. 4, 1964, pp. 188–201.

——, "A Decomposition Algorithm for Quadratic Programming," *Cahiers du Centre d'Etudes de Recherche Operationnelle*, Vol. 8, No. 2, 1966, pp. 112–131.

WILLIAMS, A. C., "A Treatment of Transportation Problems by Decomposition," *Journal of the Society for Industrial and Applied Mathematics*, Vol. 10, No. 1 (March 1962), pp. 35–48.

——, "The Status of Linear and Nonlinear Programming," paper presented at AAAS Boston Meeting, December 26–31, 1969.

WILLOUGHBY, R. A. (ed.), *Proceedings of a Symposium on Sparse Matrices*, IBM Watson Research Center, Yorktown Heights, New York, 1969.

WILSON, R., "Computation of Optimal Controls," *Journal of Mathematical Analysis and Application*, Vol. 14, No. 1 (April 1966), pp. 77–82.

WISMER, D. A. (ed.), *Optimization Methods for Large-Scale Systems*, McGraw-Hill Book Company, 1970.

WOLFE, P. AND DANTZIG, G. B., "Linear Programming in a Markov Chain," *Operations Research*, Vol. 10, No. 5 (September–October 1962), pp. 702–710.

*—— AND CUTLER, L., "Experiments in Linear Programming," in R. L. Graves and P. Wolfe (eds.), *Recent Advances in Mathematical Programming*, McGraw-Hill Book Company, New York, 1963.

*ZOUTENDIJK, G., *Methods of Feasible Directions*, Elsevier Publishing Company, Amsterdam, 1960.

ZSCHAU, E. V. W., "A Primal Decomposition Algorithm for Linear Programming," Graduate School of Business Working Paper No. 91, Stanford University, Stanford, California, January 1967.

3. DUALITY IN NONLINEAR PROGRAMMING: A SIMPLIFIED APPLICATIONS-ORIENTED DEVELOPMENT*

A. M. GEOFFRION†

Summary. The number of computational or theoretical applications of nonlinear duality theory is small compared to the number of theoretical papers on this subject over the last decade. This study attempts to rework and extend the fundamental results of convex duality theory so as to diminish the existing obstacles to successful application. New results are also given having to do with important but usually neglected questions concerning the computational solution of a program via its dual. Several applications are made to the general theory of convex systems.

The general approach is to exploit the powerful concept of a perturbation function, thus permitting simplified proofs (no conjugate functions or fixed-point theorems are needed) and useful geometric and mathematical insights. Consideration is limited to finite-dimensional spaces.

An extended summary is given in the Introduction.

CONTENTS

---

* Received by the editors September 25, 1969.

† Graduate School of Business Administration, University of California, Los Angeles, California 90024. This work was supported by the National Science Foundation under Grant GP-8740, **and by** the United States Air Force under Project RAND.

1. Introduction.

1.1. Objective. In this paper much of what is known about duality theory for nonlinear programming is reworked and extended so as to facilitate more readily computational and theoretical applications. We study the dual problem in what is probably its most satisfactory formulation, permit only assumptions that are likely to be verifiable, and attempt to establish a theory that is more versatile and general in applicability than any heretofore available.

Our methods rely upon the relatively elementary theory of convexity. No use is made of the differential calculus, general minimax theorems, or the conjugate function theory employed by most other studies in duality theory. This is made possible by fully exploiting the powerful concept of a certain *perturbation function*—the optimal value of a program as a function of perturbations of its "right-hand side." In addition to some pedagogical advantages, this approach affords deep geometrical and mathematical insights and permits a development which is tightly interwoven with optimality theory.

The resulting theory appears quite suitable for applications. Several illustrative theoretical applications are made; and some reasonable conditions are demonstrated under which the dual problem is numerically stable for the recovery of an optimal or near-optimal primal solution. A detailed preview of the results obtained is given after the canonical primal and dual programs are introduced.

1.2. The canonical primal and dual programs. The canonical *primal* problem is taken to be:

(P) $\underset{x \in X}{\text{Minimize}} \quad f(x) \quad \text{subject to} \quad g(x) \leqq 0,$

where $g(x) \triangleq (g_1(x), \cdots, g_m(x))^t$, and f and each g_i are real-valued functions defined on $X \subseteq R^n$. *It is assumed throughout that X is a nonempty convex set on which all functions are convex.*

The *dual* of (P) with respect to the g-constraints is:

(D) $\underset{u \geqq 0}{\text{Maximize}} \quad [\underset{x \in X}{\text{infimum}} \, f(x) + u^t g(x)],$

where u is an m-vector of dual variables. Note that the maximand of (D) is a concave function of u alone (even in the absence of the convexity assumptions), for it is the pointwise infimum of a collection (indexed by x) of functions linear in u.

Several other possible "duals" of (P) have been studied, some of which are discussed in § 6. All are closely related, but we believe (D) to be the most natural and useful choice for most purposes.

It is important to recognize that, given a convex program, one can dualize with respect to *any* subset of the constraints. That is, each constraint can be assigned to the g-constraints, in which case it will possess a dual variable of its own; or it can be assigned to X, in which case it will not possess a dual variable. In theoretical applications, the assignment will usually be dictated by the desired conclusion (cf. § 8); while in computational applications, the choice is usually made so that evaluating the maximand of (D) for fixed $u \geqq 0$ is significantly easier than solving (P) itself (cf. [17], [19], [29]).

1.3. Preview and summary of results. Section 2 presents the fundamental optimality and duality results as three theorems. Theorem 1 is the *optimality theorem*. Its first assertion is that if (P) has an optimal solution, then an optimal multiplier vector exists if and only if (P) has a property called *stability*, which means that the perturbation function mentioned above does not decrease infinitely steeply in any perturbation direction. Stability plays the role customarily assumed in Kuhn–Tucker type optimality theorems by some type of "constraint qualification." Because stability is necessary as well as sufficient for an optimal multiplier vector to exist, it is evidently implied by every known constraint qualification. It turns out to be a rather pleasant property to work with mathematically, and to interpret in many problems. The second assertion of the optimality theorem is that the optimal multiplier vectors are precisely the negatives of the subgradients of the perturbation function at the point of no perturbation. An immediate consequence is a rigorous interpretation of an optimal multiplier vector in terms of quasi "prices."

Theorem 2 is the customary *weak duality theorem*, which asserts that the infimal value of (P) cannot be smaller than the supremal value of (D). Although nearly trivial to show, it does have several uses. For instance, it implies that (P) must be infeasible if the supremal value of (D) is $+\infty$.

Theorem 3 is the powerful *strong duality theorem*: If (P) is stable, then (D) has an optimal solution; the optimal values of (P) and (D) are equal; the optimal solutions of (D) are essentially the optimal multiplier vectors for (P); and any optimal solution of (D) permits recovery of all optimal solutions of (P) (if such exist) as the minimizers over X of the corresponding Lagrangean function which also satisfy $g(x) \leq 0$ and the usual complementary slackness condition. Note that stability plays a central role here, just as it does in the optimality theorem. It makes (D) quite inviting as a surrogate problem for (P), and precludes the possibility of a "duality gap"—inequality between the optimal values of (P) and (D)—whose presence would render (D) useless in many, if not most, potential applications.

Section 3 gives complete proofs of these key results. The main construct is the perturbation function already mentioned. By systematically exploiting its convexity, no advanced methods or results are needed to achieve a direct and unified development of the optimality and strong duality theorems. No differentiability or even continuity assumptions need be made, and X need not be closed or open.

Section 4 develops a useful geometric portrayal of the dual problem which permits construction of simple examples to illustrate the various relationships that can obtain between (P) and (D). It also yields geometric insights which suggest some of the further theoretical results developed in the next section.

Section 5 establishes six additional theorems, numbers 4 through 9, concerning (P) and (D). *Theorem 4* asserts that the maximand of (D) has value $-\infty$ for all $u \geq 0$ if and only if the right-hand side of (P) can be perturbed so as to yield an infimal value of $-\infty$. *Theorem 5* asserts that if (P) is infeasible and yet the supremal value of (D) is finite, then some arbitrarily small perturbation of the right-hand side of (P) will restore it to feasibility. *Theorem 6* amounts to the statement that (P) is stable if it satisfies Slater's qualification that there exist a point $x^0 \in X$

such that $g_i(x^0) < 0$, $i = 1, \cdots, m$. The next result, called the *continuity theorem*, gives a key necessary and sufficient condition for the optimal values of (P) and (D) to be equal: the perturbation function must be lower semicontinuous at the origin. (Stability is just a sufficient condition, in view of the strong duality theorem, unless further assumptions are made.) *Theorem* 8 provides useful sufficient conditions for the perturbation function to be lower semicontinuous at the origin; and, therefore, for a duality gap to be impossible: f and g continuous, X closed, the infimal value of (P) finite, and the set of ε-optimal solutions nonempty and bounded for some $\varepsilon > 0$. The final result of § 5, called the *converse duality theorem*, is an important companion to the strong duality theorem. It requires X to be closed and f and g to be continuous on X; and asserts that if (D) has an optimal solution and the corresponding Lagrangean function has a unique minimizer x^* over X which is also feasible in (P), then (P) is stable and x^* is the unique optimal solution. Actually, as the discussion indicates, the hypothesis concerning the uniqueness of the minimizer of the Lagrangean can be weakened.

Section 6 examines the relationships between the results of §§ 1 to 5 and previous work. After discussing the specialization to the completely linear case, a detailed comparison is made to several key papers representative of the major approaches previously applied to nonlinear duality theory. These are: Dorn [8], which exploits the special properties of quadratic programs; Wolfe [34], which applies the differential calculus and the classical results of Kuhn and Tucker; Stoer [28] and Mangasarian and Pontstein [25], which apply general minimax theorems; and Rockafellar [26], which applies conjugate convex function theory. This comparison is favorable to the methods and results of the present study.

Section 7 discusses numerical considerations of interest if (D) is to be used computationally to solve (P). Such questions have been almost totally ignored in previous studies, but must be examined if nonlinear duality theory is to be applied computationally. After first indicating some of the pitfalls stability precludes, we briefly survey the two main approaches that have been followed for optimizing (D); and, subsequently, the main topic of whether and how an optimal or near-optimal solution of a stable program can be obtained from an optimal or near-optimal solution of its dual. A result is given in *Theorem* 10 from which it follows that no particular numerical difficulties exist in solving (P), provided that an exactly optimal solution of (D) can be found. However, if only a sequence converging to an optimal solution u^* of (D) can be found, the situation appears to turn on whether the Lagrangean function corresponding to u^* has a unique minimizer over X. If it does, *Theorem* 11 shows that the situation is manageable, at least when X is compact and f and g are continuous on X. Otherwise, the situation can be quite difficult, as demonstrated by an example.

Section 8 makes the point that nonlinear duality theory can be used to prove many results in the theory of convex systems which do not appear to involve optimization at all; just as linear duality theory can be used to prove results concerning systems of linear equalities and inequalities. This provides an easy and unified approach to a substantial body of theorems. The possibilities are illustrated by presenting new proofs for three theorems. The first is a separation theorem for disjoint convex sets; the second a characterization in terms of supporting half-spaces for a certain class of convex sets generated by projection; and the third a fundamental property of a system of inconsistent convex inequalities.

Finally, in § 9 we indicate a few of the significant areas in which further work remains to be done.

1.4. Notation. The notation employed is standard. We follow the convention that all vectors are columnar unless transposed (e.g., u^t).

2. Fundamental results. After establishing some basic definitions, we state and discuss three fundamental results: the optimality theorem, weak duality theorem, and strong duality theorem. The proofs of the first and third results are deferred to § 3.

2.1. Definitions.

DEFINITION 1. The *optimal value of* (P) is the infimum of $f(x)$ subject to $x \in X$ and $g(x) \leq 0$. The *optimal value of* (D) is the supremum of its maximand subject to $u \geq 0$.

Problems (P) and (D) always have optimal values (possibly $\pm \infty$) whether or not they have optimal solutions—that is, whether or not there exist feasible solutions achieving these values—provided we invoke the customary convention that an infimum (supremum) taken over an empty set is $+\infty(-\infty)$.

DEFINITION 2. A vector u is said to be *essentially infeasible in* (D) if it yields a value of $-\infty$ for the maximand of (D). If every $u \geq 0$ is essentially infeasible in (D), then (D) itself is said to be *essentially infeasible*. If (D) is not essentially infeasible, it is said to be *essentially feasible*.

The motivation for this definition is obvious: a vector $u \geq 0$ is useless in (D) if it leads to an "infinitely bad" value of the maximand.

DEFINITION 3. A pair (x, u) is said to satisfy the *optimality conditions* for (P) if

(i) x minimizes $f + u^t g$ over X,
(ii) $u^t g(x) = 0$,
(iii) $u \geq 0$,
(iv) $g(x) \leq 0$.

A vector u is said to be an *optimal multiplier vector* for (P) if (x, u) satisfies the optimality conditions for some x.

An optimal multiplier vector is sometimes referred to as a "generalized Lagrange multiplier vector," or a vector of "dual variables" or "dual prices."

It is easy to verify that a pair (x, u) satisfies the optimality conditions only if x is optimal in (P). Thus the existence of an optimal multiplier vector presupposes the existence of an optimal solution of (P). The converse, of course, is not true without qualification. It also can be verified that if u is an optimal multiplier vector, then (x, u) satisfies the optimality conditions for *every* optimal solution of (P). Thus an optimal multiplier vector is truly associated with (P)—more precisely, with the optimal solution set of (P)—rather than with any particular optimal solution. On this point the traditional custom of defining an optimal multiplier vector in terms of a particular optimal solution of (P) is misleading, although it is equivalent to the definition used here.

It is perhaps worthwhile to remind the reader that the optimality conditions are equivalent to a constrained saddle point of the Lagrangean function. Specifically, one can verify that (x^*, u^*) satisfies conditions (i)–(iv) if and only if $u^* \geq 0$, $x^* \in X$, and

$$f(x^*) + u^t g(x^*) \leq f(x^*) + (u^*)^t g(x^*) \leq f(x) + (u^*)^t g(x)$$

for all $u \geq 0$ and $x \in X$. Another equivalent rendering of the optimality conditions, one which gives a glimpse of developments to come, is this: (x^*, u^*) satisfies the optimality conditions if and only if x^* is optimal in (P), u^* is optimal in (D), and the optimal values of (P) and (D) are equal.

These remarks on Definition 3 do not depend in any way on the convexity assumptions. The demonstrations are straightforward.

DEFINITION 4. The *perturbation function* $v(\cdot)$ associated with (P) is defined on R^m as

$$v(y) \triangleq \operatorname*{infimum}_{x \in X} \{ f(x) \text{ subject to } g(x) \leq y \},$$

where y is called the *perturbation vector*.

The perturbation function is convex (Lemma 1) and is the fundamental construct used to derive the relationship between (P) and (D). Evidently $v(0)$ is the optimal value of (P). Values of v at points other than the origin are also of intrinsic interest in connection with sensitivity analysis and parametric studies of (P).

DEFINITION 5. Let $\bar{y}$ be a point at which v is finite. An m-vector $\bar{\gamma}$ is said to be a *subgradient* of v at $\bar{y}$ if [1]

$$v(y) \geq v(\bar{y}) + \bar{\gamma}^t(y - \bar{y}) \quad \text{for all } y.$$

Subgradients generalize the concept of a gradient and are a technical necessity since v is usually not everywhere differentiable. Their role is made even more important by the fact that they turn out to be the negatives of the optimal multiplier vectors (Lemma 3). An important criterion for their existence (Lemma 2) is given by the property in the next definition (cf. Gale [15]).

DEFINITION 6. (P) is said to be *stable* if $v(0)$ is finite and there exists a scalar $M > 0$ such that

$$\frac{v(0) - v(y)}{\|y\|} \leq M \quad \text{for all } y \neq 0.$$

Stability is an easy property to understand intuitively. We shall see that it is implied by all known constraint qualifications for (P). It can be interpreted as a Lipschitz condition on the function v. If it fails to hold, then the ratio of improvement in the infimal value of (P) to the amount of perturbation can be made as large as desired (the particular norm $\| \cdot \|$ used to measure the amount of perturbation is immaterial). This is also true in the marginal sense, that is, with the

---

[1] The direction of this inequality would be reversed if v were concave rather than convex.

perturbations made as small as desired, as follows from the convexity of v. A consequence of this observation is that the following alternative definition of stability (used by Rockafellar) is equivalent to the one above.

DEFINITION 6'. (P) is said to be *stable* if v is finite at 0 and does not decrease infinitely steeply in any perturbation direction; that is, if

$$\lim_{\theta \to 0+} \left[\frac{v(0) - v(\theta y)}{\theta \|y\|} \right] < \infty \quad \text{for all } y \neq 0.$$

The limit defined is the negative of the directional derivative of v in the perturbation direction y ($\pm \infty$ are allowed as limits).

2.2. Optimality. Although the main focus of this study is duality theory, an inevitable by-product of the present approach is what must surely be near the ultimate of Kuhn–Tucker type optimality theorems. (Cf. Gale [15, Theorem 3] and Rockafellar [26].) The following theorem also gives a key characterization of optimal multiplier vectors.

THEOREM 1 (Optimality). *Assume that* (P) *has an optimal solution. Then an optimal multiplier vector exists if and only if* (P) *is stable; and u is an optimal multiplier vector for* (P) *if and only if* $(-u)$ *is a subgradient of v at* $y = 0$.

Part of the content of this theorem is the result that, if x^* is an optimal solution of (P), and (P) is stable, then there exists a vector u^* such that (x^*, u^*) satisfies the optimality conditions for (P). It is well known that some qualification of (P) is needed for this result to hold; stability plays the role of such a qualification here. It bears emphasizing, however, that the theorem reveals stability to be not only a sufficient qualification for this purpose, but also a *necessary* one. Thus, stability is implied by every "constraint qualification" ever used to prove the necessity of the optimality conditions. For example, Slater's constraint qualification that there exist a point $x^0 \in X$ such that $g_i(x^0) < 0$ for all i implies stability, as does the original Kuhn–Tucker constraint qualification. For a discussion of these and many other "classical" qualifications, see Mangasarian [24].

It is striking that none of the classical qualifications emphasizes the role of the objective function, although a careful reading reveals that each requires the objective function to be within a particular general class (e.g., defined on all of R^n, or differentiable on an open set containing the feasible region). Of course, if the objective function is sufficiently well-behaved, (P) will be stable no matter how poorly behaved the constraints are (e.g., if f is constant on X then (P) is obviously stable for any constraint set as long as it is feasible). On the other hand, it is possible for an objective function to be so poorly behaved that (P) is unstable even if all constraints are linear. For example [15], put $n = m = 1$, $f(x) = -\sqrt{x}$, $X = \{x : x \geq 0\}$, $g_1(x) = x$; then by perturbing the right-hand side positively, the ratio

$$\frac{0 - (-\sqrt{y})}{|y|} = \frac{1}{\sqrt{y}}$$

in Definition 6 can be made as large as desired by making the perturbation amount

y sufficiently small. The obvious trouble with the objective function is that it has infinite steepness at $x = 0$.

One further useful and somewhat surprising observation on the concept of stability is in order. Namely, to verify stability it is actually necessary and sufficient to consider only a *one-dimensional* choice of y: (P) is stable if and only if $v(0)$ is finite and there exists a scalar $M > 0$ such that

$$\frac{v(0) - v(\xi, \cdots, \xi)}{\xi} \leq M \quad \text{for all } \xi > 0.$$

The proof makes use of the fact that stability according to Definition 6 does not depend on which particular norm is used (this follows from the fact that if $\| \cdot \|_1$ and $\| \cdot \|_2$ are two different norms on R^m, then there exists a scalar $r > 0$ such that $(\|y\|_1 / \|y\|_2) \geq r$ for all nonzero $y \in R^m$), so that the Chebyshev norm

$$\|y\|_T \equiv \text{maximum } \{|y_1|, \cdots, |y_m|\}$$

can be used in Definition 6. Thus stability of (P) implies the above inequality simply by taking y of the form $(\xi, \cdots, \xi)$. Conversely, the above inequality implies that (P) is stable because

$$\frac{v(0) - v(y)}{\|y\|_T} \leq \frac{v(0) - v(\|y\|_T, \cdots, \|y\|_T)}{\|y\|_T}, \qquad \text{all } y \neq 0,$$

by the fact that v is a nonincreasing function.

The second part of Theorem 1 gives a characterization of optimal multiplier vectors useful for sensitivity analysis and other purposes. Suppose that u^* is any optimal multiplier vector of (P) and a perturbation in the direction y' is contemplated; that is, we are interested in the perturbed problem:

$$\underset{x \in X}{\text{Minimize }} f(x) \quad \text{subject to} \quad g(x) \leq \theta y'$$

for $\theta \geq 0$. From Theorem 1,

$$v(\theta y') \geq v(0) - \theta (u^*)^t y' \quad \text{for all } \theta \geq 0.$$

We thus have a lower bound on the optimal value of the perturbed problem, and by taking limits we can obtain a bound on the right-hand derivative of $v(\theta y')$ at $\theta = 0$;

$$\frac{d^+ v(\theta y')}{d\theta} \geq -(u^*)^t y'.$$

In particular, with y' equal to the jth unit vector, we obtain the well-known result that $-u_j^*$ is a lower bound on the marginal rate of change of the optimal value of (P) with respect to an increase in the right-hand side of the jth constraint. All of this follows from the fact that any optimal multiplier vector is the negative of a subgradient of v at 0. The converse of this is also significant, as it follows that the set of all subgradients—and hence all directional derivatives of v—can be characterized in terms of the set of all optimal multiplier vectors for (P). Further details and an application to the optimization of multidivisional systems are given elsewhere [16, § 4.2].

2.3. Duality. We begin with a very easy result that has several useful consequences.

THEOREM 2 (Weak duality). *If $\bar{x}$ is feasible in* (P) *and $\bar{u}$ is feasible in* (D), *then the objective function of* (P) *evaluated at $\bar{x}$ is not less than the objective function of* (D) *evaluated at $\bar{u}$.*

To demonstrate this result, one need only write the obvious inequalities

$$\text{infimum } \{f(x) + \bar{u}'g(x)|x \in X\} \leq f(\bar{x}) + \bar{u}'g(\bar{x}) \leq f(\bar{x}).$$

The convexity assumptions are not needed.

One consequence is that any feasible solution of (D) provides a lower bound on the optimal value of (P); and any feasible solution of (P) provides an upper bound on the optimal value of (D). This can be useful in establishing termination or error-control criteria when devising computational algorithms addressed to (P) or (D); if at some iteration feasible solutions are available to both (P) and (D) that are "close" to one another in value, then they must be "close" to being optimal in their respective problems. In Theorem 3 we shall see that there exist feasible solutions to (P) and (D) that are as close to one another in value as desired, provided only that (P) is stable.

From Theorem 2, it also follows that (D) must be essentially infeasible if the optimal value of (P) is $-\infty$ and, similarly, (P) must be infeasible if the optimal value of (D) is $+\infty$.

THEOREM 3 (Strong duality). *If* (P) *is stable, then*

(a) (D) *has an optimal solution,*

(b) *the optimal values of* (P) *and* (D) *are equal,*

(c) u^* *is an optimal solution of* (D) *if and only if* $-u^*$ *is a subgradient of v at* $y = 0$,

(d) *every optimal solution u^* of* (D) *characterizes the set of all optimal solutions (if any) of* (P) *as the minimizers of $f + (u^*)'g$ over X which also satisfy the feasibility condition $g(x) \leq 0$ and the complementary slackness condition $(u^*)'g(x) = 0$.*

Conclusions (a) and (d) justify taking a dual approach to the solution of (P). It is perhaps surprising that *all* optimal solutions of (P) can be found from any *single* optimal solution of (D). Another way of phrasing (d) would be to say that if u^* is optimal in (D), then x is optimal in (P) if and only if (x, u^*) satisfies optimality conditions (i), (ii) and (iv) (see Definition 3). In § 7 we shall take up at some length the matter of approaching the computational solution of (P) via its dual.

Conclusion (b) precludes the existence of what is often referred to as a *duality gap* between the optimal values of (P) and (D). Most applications of nonlinear duality theory require that there be no duality gap (e.g., see § 8 and [18]).

Conclusion (c) reveals the connection between the set of optimal solutions of (D) and the perturbation function. If (P) has an optimal solution as well as the property of stability, then, using Theorem 1, we obtain an alternative interpretation of the optimal solution set of (D): it is precisely the set of optimal multiplier vectors for (P).

It is perhaps worth noting that Lemma 4 in the next section shows that conclusion (c) holds under a slightly weaker assumption than stability, namely, when $v(0)$ is finite and the optimal values of (P) and (D) are equal.

Additional results concerning (P) and (D) are given in § 5. Relationships to known results are taken up in § 6.

3. Proof of the optimality and strong duality theorems. It is convenient to subdivide the proofs of Theorems 1 and 3 into five lemmas.

It will be necessary to refer to the set Y of all vectors y for which the perturbed problem is feasible:

$$Y \triangleq \{y \in R^m : g(x) \leqq y \text{ for some } x \in X\}.$$

Obviously, $v(y) = \infty$ if and only if $y \notin Y$.

The first lemma establishes that the perturbation function is convex on Y. This well-known result is the cornerstone of the entire development. It also is obvious that v is nonincreasing.

LEMMA 1. *Y is a convex set, and v is a convex function on Y.*

Proof. The convexity of Y follows directly from the convexity of X and the convexity of g. Since $-\infty$ is permitted as a value for v on Y, the appropriate definition of convexity for v is in terms of its epigraph [26]: $\{(y, \mu) \in R^{m+1} : y \in Y$ and $\mu \geqq v(y)\}$. Let (y^0, μ^0) and (y', μ') be arbitrary points in this set, and let θ be an arbitrary scalar between 0 and 1. Define $\bar{\theta} = 1 - \theta$. Then

$$v(\theta y^0 + \bar{\theta} y') = \inf_{x^0, x' \in X} f(\theta x^0 + \bar{\theta} x') \quad \text{subject to } g(\theta x^0 + \bar{\theta} x') \leqq \theta y^0 + \bar{\theta} y'$$

$$\leqq \inf_{x^0, x' \in X} f(\theta x^0 + \bar{\theta} x') \quad \text{subject to } g(x^0) \leqq y^0, g(x') \leqq y'$$

$$\leqq \inf_{x^0, x' \in X} \theta f(x^0) + \bar{\theta} f(x') \quad \text{subject to } g(x^0) \leqq y^0, g(x') \leqq y'$$

$$= \theta v(y^0) + \bar{\theta} v(y') \leqq \theta \mu^0 + \bar{\theta} \mu',$$

where the equality or inequality relations follow, respectively, from the convexity of X, the convexity of g, the convexity of f, separability in x^0 and x', and the definitions of μ^0 and μ'. Thus the point $\theta(y^0, \mu^0) + \bar{\theta}(y', \mu')$ is in the epigraph of v, and so v must be convex on Y. This completes the proof.

Many properties of v follow directly from its convexity. For example, v must be continuous on the interior of any set on which it is finite; it must have value $-\infty$ everywhere on the interior of Y if it is $-\infty$ anywhere; its directional derivative (see Definition 6') must exist in every direction at every point where it is finite; it must have a subgradient at every interior point of Y at which it is finite; and it must be differentiable at a given point in Y if and only if it has a unique subgradient there. These properties hold, not only for v, but for any convex function (see, e.g., [11] or [26, § 2]).

The following property is an important criterion for the existence of a subgradient of a convex function at a point where it is finite. We offer a proof to keep the development self-contained; the method of proof is due to Gale [15].

LEMMA 2. *Let $\phi(\cdot)$ be a convex function on a convex set $Y \subseteq R^m$ taking values in $R \cup \{-\infty\}$. Let $\|\cdot\|$ be any norm on R^m, and let $\bar{y}$ be a point at which ϕ is finite. Then ϕ has a subgradient at $\bar{y} \in Y$ if and only if there exists a positive scalar M such that*

$$\frac{\phi(\bar{y}) - \phi(y)}{\|y - \bar{y}\|} \leqq M \quad \text{for all } y \in Y \quad \text{such that } y \neq \bar{y}.$$

Proof. Suppose that ϕ has a subgradient $\bar{\gamma}$ at $\bar{y} \in Y$; i.e.,

$$\phi(y) \geq \phi(\bar{y}) + \bar{\gamma}'(y - \bar{y}) \quad \text{for all } y \in Y.$$

Then $\phi(y) > -\infty$ on Y, and upon rearranging and dividing by $\|y - \bar{y}\|$ (we shall use the Euclidean norm, although any norm will do) we obtain

$$\frac{\phi(\bar{y}) - \phi(y)}{\|y - \bar{y}\|} \leq \frac{-\bar{\gamma}'(y - \bar{y})}{\|y - \bar{y}\|} \quad \text{for all } y \in Y \quad \text{such that } y \neq \bar{y}.$$

Since the right-hand side does not exceed $\|\bar{\gamma}\|$ for any y, we obtain the desired inequality with $M = \|\bar{\gamma}\|$. (If $\|\bar{\gamma}\| = 0$, the desired inequality holds with M equal to any positive number.)

Now suppose that there exists a positive scalar M such that the stated inequality holds. We must show that there exists a subgradient of ϕ at $\bar{y}$. Since $\phi(y) > -\infty$ on Y, we may define the sets

$$\Phi \triangleq \{(y, z) \in R^{m+1} : y \in Y \text{ and } \phi(\bar{y}) - \phi(y) \geq z\},$$

$$\Psi \triangleq \{(y, z) \in R^{m+1} : M\|y - \bar{y}\| < z\}.$$

It is easy to see that Φ and Ψ are convex sets, that $\Phi \cap \Psi$ is empty, and that Ψ is open. From elementary results on the separation of nonintersecting convex sets, it follows that Φ and Ψ can be separated by a hyperplane that does not intersect Ψ; that is, there exist an m-vector γ and scalars ρ and α such that

$$\gamma'y + \rho z \geq \alpha \quad \text{for } (y, z) \in \Phi, \qquad \gamma'y + \rho z < \alpha \quad \text{for } (y, z) \in \Psi,$$

Now $(\bar{y}, 0) \in \Phi$, so $\gamma'\bar{y} \geq \alpha$. Actually $\gamma'\bar{y} = \alpha$, for $\gamma'\bar{y} \leq \alpha$ follows from the second inequality and the fact that $(\bar{y}, \varepsilon) \in \Psi$ for all $\varepsilon > 0$. Thus the inequalities become

$$\gamma'(y - \bar{y}) + \rho z \geq 0 \quad \text{for } (y, z) \in \Phi,$$

$$\gamma'(y - \bar{y}) + \rho z < 0 \quad \text{for } (y, z) \in \Psi.$$

Since $(\bar{y}, 1) \in \Psi$, we have $\rho < 0$. Put $\tilde{\gamma} = -\gamma'/\rho$. Then the first inequality becomes $\tilde{\gamma}'(y - \bar{y}) \geq z$ whenever $(y, z) \in \Phi$, that is, whenever $y \in Y$ and $\phi(\bar{y}) - \phi(y) \geq z$. Putting $z = \phi(\bar{y}) - \phi(y)$, we have

$$\tilde{\gamma}'(y - \bar{y}) \geq \phi(\bar{y}) - \phi(y) \quad \text{whenever } y \in Y,$$

which says precisely that $-\tilde{\gamma}$ is a subgradient of ϕ at $y = \bar{y}$. This completes the proof.

The following known result is equivalent: a convex function has a subgradient at a given point where it is finite if and only if its directional derivative is not $-\infty$ in any direction (cf. [11, p. 84], [26, p. 408]). This result would be used in place of Lemma 2 if Definition 6' were used in place of Definition 6.

The next lemma establishes a crucially important alternative interpretation of optimal multiplier vectors.

LEMMA 3. *If* (P) *has an optimal solution, then* u *is an optimal multiplier vector for* (P) *if and only if* $-u$ *is a subgradient of* v *at* $y = 0$.

Proof. Suppose that u^* is an optimal multiplier vector for (P). Then there is a vector x^* such that (x^*, u^*) satisfies the optimality conditions for (P). From optimality conditions (i) and (ii) we have $f(x) \geq f(x^*) - (u^*)'g(x)$ for all $x \in X$.

It follows, using (iii), that for each point y in Y,

$$f(x) \geq f(x^*) - (u^*)^t y \quad \text{for all } x \in X \quad \text{such that } g(x) \leq y.$$

Taking the infimum of the left-hand side of this inequality over the indicated values of x yields

$$v(y) \geq f(x^*) + (-u^*)^t y \quad \text{for all } y \in Y.$$

It is now evident that $-u^*$ satisfies the definition of a subgradient of v at 0 (since $f(x^*) = v(0)$ and $v(y) = \infty$ for $y \notin Y$). This completes the first half of the proof.

Now let $-u^*$ be any subgradient of v at 0. We shall show that (x^*, u^*) satisfies optimality conditions (i)–(iv), where x^* is any optimal solution of (P). Condition (iv) is immediate. The demonstration of the remaining conditions follows easily from the definitional inequality of $-u^*$, namely,

$$v(y) \geq v(0) - (u^*)^t y \quad \text{for all } y.$$

To establish (iii), put $y = e_j$, the jth unit m-vector (all components of e_j are 0 except for the jth, which is 1). Then $v(e_j) \geq v(0) - u_j^*$, or $u_j^* \geq v(0) - v(e_j)$. Since v is obviously a nonincreasing function of y, we have $v(0) \geq v(e_j)$ and therefore $u_j^* \geq 0$. Condition (ii) is established in a similar manner by putting $y = g(x^*)$. This yields $(u^*)^t g(x^*) \geq v(0) - v(g(x^*)) = 0$, where the equality follows from the fact that decreasing the right-hand side of (P) to $g(x^*)$ will not destroy the optimality of x^*. Hence $(u^*)^t g(x^*) \geq 0$. But the reverse inequality also holds by (iii) and the feasibility of x^* in (P), and so (ii) must hold. To establish condition (i), put $y = g(x)$, where x is any point in X. Then

$$v(g(x)) \geq v(0) - (u^*)^t g(x) \quad \text{for all } x \in X.$$

Since $f(x) \geq v(g(x))$ for all $x \in X$ and $v(0) = f(x^*)$, we have

$$f(x) \geq f(x^*) - (u^*)^t g(x) \quad \text{for all } x \in X.$$

In view of (ii), this is precisely condition (i). This completes the proof.

We now have all the ingredients necessary for the optimality theorem.

Proof of Theorem 1. The perturbation function is convex on Y, by Lemma 1, and is finite at $y = 0$ because (P) is assumed to possess an optimal solution. By Lemma 2, therefore, v has a subgradient at $y = 0$ if and only if (P) is stable. But subgradients of v at $y = 0$ and optimal multiplier vectors for (P) are negatives of one another by Lemma 3. The conclusions of Theorem 1 are now at hand.

It is worth digressing for a moment to note that the hypotheses of Lemma 3 and Theorem 1 could be weakened slightly if the definition of an optimal multiplier vector were generalized so that it no longer presupposes the existence of an optimal solution of (P). In particular, the conclusions of Lemma 3 and Theorem 1 hold even if (P) does not have an optimal solution, provided that $v(0)$ is finite and the concept of an optimal multiplier vector is redefined as follows. A point u is said to be a *generalized optimal multiplier* vector if for every scalar $\varepsilon > 0$ there exists a point x_ε such that (x_ε, u) satisfies the *ε-optimality conditions*: x is an ε-optimal minimizer of $f + u^t g$ over X, $u^t g(x) \geq \varepsilon$, $u \geq 0$, and $g(x) \leq 0$. The necessary modification of the proof of Lemma 3 is straightforward.

The entire development of this paper could be carried out in terms of this more general concept of an optimal multiplier vector. The optimality conditions for (P), for example, would be stated in terms of a pair $(\langle x^v \rangle, u)$ in which the first member is a sequence rather than a single point. Such a pair would be said to satisfy the *generalized optimality conditions* for (P) if, for some nonnegative sequence $\langle \varepsilon^v \rangle$ converging to 0, for each v the pair (x^v, u) satisfies the ε^v-optimality conditions. This generalization of the traditional optimality conditions given in Definition 3 seems to be the most natural one, when the existence of an optimal solution of (P) is in question. It can be shown that if u is a generalized optimal multiplier vector then $(\langle x^v \rangle, u)$ satisfies the generalized optimality conditions if and only if $\langle x^v \rangle$ is a sequence of feasible solutions of (P) converging in value to $v(0)$.

Although such a development might well be advantageous for some purposes, we elect not to pursue it here.[2]

The next lemma establishes (in view of the previous one) the connection between optimal multiplier vectors and solutions to the dual problem.

LEMMA 4. *Let $v(0)$ be finite. Then u is an optimal solution of* (D) *and the optimal values of* (P) *and* (D) *are equal if and only if* $-u$ *is a subgradient of v at $y = 0$.*

Proof. First we demonstrate the "if" part of the lemma. Let $(-u)$ be a subgradient of v at $y = 0$; that is, let u satisfy

$$v(y) \geqq v(0) - u^t(y - 0) \quad \text{for all } y.$$

The proof of Lemma 3 shows that $u \geqq 0$ follows from this inequality. Thus u is feasible in (D). Substituting $g(x)$ for y, and noting that $f(x) \geqq v(g(x))$ holds for all $x \in X$ yields

$$f(x) + u^t g(x) \geqq v(0) \quad \text{for all } x \in X.$$

Taking the infimum over $x \in X$, we obtain

$$\text{infimum}_{x \in X} \{ f(x) + u^t g(x) \} \geqq v(0).$$

It now follows from the weak duality theorem that u must actually be an optimal solution of (D), and that the optimal value of (D) equals $v(0)$. This completes the first part of the proof.

To demonstrate the "only if" part of the lemma, let u be an optimal solution of (D). By assumption,

$$[\text{infimum}_{x \in X} f(x) + u^t g(x)] = v(0).$$

Since $u^t g(x) \leqq u^t y$ for all $x \in X$ and y such that $g(x) \leqq y$ (remember that $u \geqq 0$), it follows that

$$f(x) + u^t y \geqq v(0)$$

for all $x \in X$ and y such that $g(x) \leqq y$. For each $y \in Y$, we may take the infimum

---

[2] K. O. Kortanek has pointed out in a private communication (Dec. 8, 1969) that the concept of an optimal multiplier vector can be generalized still further by considering a sequence $\langle u^v \rangle$ of m-vectors and appropriately defining "asymptotic" optimality conditions (cf. [21]). This would permit a kind of optimality theory for many unstable problems (such as the example in § 2.2), although interpretations in terms of subgradients of v at $y = 0$ are obviously not possible.

of this inequality over $x \in X$ such that $g(x) \leq y$ to obtain

$$v(y) - (-u)'y \geq v(0) \quad \text{for } y \in Y.$$

This inequality holds outside of Y as well, since $v(y) = \infty$ there. Thus $(-u)$ satisfies the definition of a subgradient of v at 0, and the proof is complete.

The final lemma characterizes the (possibly empty) optimal solution set of (P).

LEMMA 5. *Assume that v is finite at $y = 0$ and that γ is a subgradient at this point. Then x^* is an optimal solution of* (P) *if and only if $(x^*, -\gamma)$ satisfies optimality conditions* (i), (ii) *and* (iv) *for* (P).

Proof. Let x^* be an optimal solution of (P). By Lemma 3, $(-\gamma)$ must be an optimal multiplier vector for (P); and so $(x^*, -\gamma)$ must satisfy the optimality conditions for (P) (see the discussion following Definition 3). This proves the "only if" part of the conclusion.

Now let $(x^*, -\gamma)$ satisfy (i), (ii) and (iv). The proof of Lemma 4 shows that $-\gamma \geq 0$, and so (iii) is also satisfied. Hence x^* must be an optimal solution of (P). This completes the proof.

We are now able to prove the strong duality theorem.

Proof of Theorem 3. Since (P) is stable, $v(0)$ is finite and we conclude from Lemmas 1 and 2 that v has a subgradient at $y = 0$. Parts (a), (b) and (c) of the theorem now follow immediately from Lemma 4. Part (d) follows immediately from Lemma 5 with the help of part (c).

4. Geometrical interpretations and examples. It is easy to give a useful geometric portrayal of the dual problem (cf. [20], [23, p. 223], [31]). This yields insight into the content of the definitions and theorems of § 2, permits construction of various pathological examples, and even suggests a number of additional theoretical results. We need to consider in detail only the case $m = 1$.

The geometric interpretation focuses on the image of X under f and g, that is, on the image set

$$I \triangleq \{(z_1, z_2) \in R^2 : z_1 = g(x) \text{ and } z_2 = f(x) \text{ for some } x \in X\}.$$

Figure 1 illustrates a typical problem in which x is a scalar variable. The point P^* is obviously the image of the optimal solution of problem (P); that is,

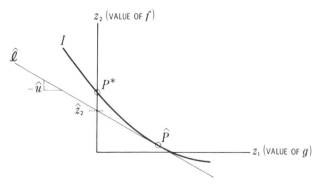

FIG. 1

$P^* = (g(x^*), f(x^*))$, where x^* minimizes f subject to $g(x) \leqq 0$ and $x \in X$. Thus the geometric interpretation of problem (P) is obvious: Find the point in I which minimizes z_2 subject to $z_1 \leqq 0$.

Consider now a particular value $\hat{u} \geqq 0$ for the scalar variable of (D). To evaluate the maximand of (D) at $\hat{u}$ one must minimize $f + \hat{u}g$ over X. This is the same as minimizing $z_2 + \hat{u}z_1$ subject to $(z_1, z_2) \in I$; as the line $z_2 + \hat{u}z_1 = \text{const.}$ has slope $-\hat{u}$ in Fig. 1, we see that evaluating the maximand of (D) at $\hat{u}$ amounts to finding the lowest line with slope $-\hat{u}$ which intersects I. This leads to the line $\hat{\ell}$ tangent to I at $\hat{P}$, pictured in Fig. 1. The point $\hat{P}$ is the image of the minimizer of $f + \hat{u}g$ over X. The minimum value of $f + \hat{u}g$ is the value of z_2 where $\hat{\ell}$ intercepts the ordinate, namely, $\hat{z}_2$ in Fig. 1 (since $(0, \hat{z}_2) \in \hat{\ell}$). The geometric interpretation of (D) is now apparent: Find that value of $\hat{u}$ which defines the slope of a line tangent to I intersecting the ordinate at the highest possible value. Or, more loosely, choose $\hat{u}$ to maximize $\hat{z}_2$. In Fig. 1, this leads to a value of u which defines a line tangent to I at P^*.

The geometric interpretation of (P) and (D) helps to clarify the content of Theorems 1, 2 and 3. The problem pictured in Fig. 1, for example, is obviously stable. In the neighborhood of $y = 0$, $v(y)$ is just the z_2-coordinate of I when z_1 equals y; and this coordinate does not decrease infinitely steeply as y deviates from 0. The subgradient of v at $y = 0$ is precisely the slope of the line tangent to I at P^*, which from Definition 3 is seen to be the negative of the optimal multiplier vector. This verifies the conclusion of Theorem 1 for this example. The geometrical verification of Theorems 2 and 3 is so easy as not to require comment here.

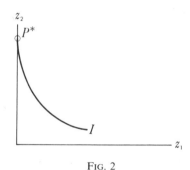

Fig. 2

An example of an unstable problem is given in Fig. 2, in which I is tangent to the ordinate at the point P^*. The value of v decreases infinitely steeply as y begins to increase above 0, and so there can be no subgradient at $y = 0$. The only line tangent to I at P^* is vertical. This checks with Theorem 1. Theorem 2 obviously holds, and Theorem 3 does not apply. The dual has optimal value equal to that of the primal, but no finite u achieves it.

This concludes the discussion of the geometrical interpretation of the definitions and theorems when $m = 1$. The generalization for $m > 1$ is conceptually straightforward, although it may be helpful to think in terms of the following convex set instead of I itself:

$$I^+ \triangleq \{(z_1, \cdots, z_{m+1}) \in R^{m+1} : z_i \geqq g_i(x), i = 1, \cdots, m,$$

$$\text{and } z_{m+1} \geqq f(x) \text{ for some } x \in X\}.$$

Using I^+ in place of I does not change in any way the geometrical interpretations given for (P) or (D). The line $\hat{\ell}$ now becomes, for each choice of $\hat{u}$, a supporting hyperplane (when $m > 1$) of I^+.

Now we shall put our geometrical insight to work by constructing enough pathological examples to show that the cases allowed by Theorems 1, 2 and 3 can actually occur. These examples are displayed on Diag. 1. The image set I is portrayed in each case. A dashed line indicates a missing boundary, and an arrowhead indicates that I continues indefinitely in the given direction. Although the examples are given geometrically, it is easy to write down corresponding expressions for $X, f,$ and g. Only a single variable is needed in Examples 1, 6, 9 and 10. In Examples 1, 6 and 9, simply identify x with z_1; let $g(x) = x$; $f(x) = z_2(x)$, where $z_2(z_1)$ is the z_2-coordinate of I for a given value of z_1; and let X be the interval of z_1 values assumed by points in I. In Example 9, for instance, the corresponding problem (P) might be:

$$\underset{1 \leqq x \leqq 3}{\text{Minimize}} \quad x \quad \text{subject to} \quad x \leqq 0.$$

For Example 10, one may put $f(x) = -x$, $g(x) \equiv +1$, and $X = [-1, \infty)$. The remaining examples require two variables: identify x_1 with z_1 and x_2 with z_2, let $g(x_1, x_2) = x_1$ and $f(x_1, x_2) = x_2$; and put X equal to I. In Example 3, for instance, we obtain for (P) the problem:

$$\underset{(x_1, x_2) \in X}{\text{Minimize}} \quad x_2 \quad \text{subject to} \quad x_1 \leqq 0,$$

where

$$X \triangleq \{(x_1, x_2) : 0 \leqq x_1 \leqq 2, 1 < x_2 \leqq 4, \text{ and } x_2 \geqq 3 \text{ if } x_1 = 0\}.$$

In Diag. 1, we have not distinguished whether or not (P) has an optimal solution when $v(0)$ is finite. There does exist an optimum solution of (P) in each of Examples 1–5; it is denoted in each case by a heavy dot. It is easy to modify each of these examples so that no optimum solution exists: simply delete the dot from I in Examples 2–5, and delete the dot and the part of I to its right in Example 1. This shows that the cases allowed for (P) in Diag. 1 when its optimal value is finite can occur either with or without the existence of an optimal solution for (P).

5. Additional results. The geometrical insights offered in the previous section, and particularly the geometric examples of Diag. 1, suggest a number of further results concerning the relation between (P) and (D). In this section we prove several results useful in ascertaining when certain of the cases represented by Examples 2–10 cannot occur. We also present a converse duality theorem which sharpens part (d) of the strong duality theorem.

5.1. Theorems 4–6. The first result suggested by Diag. 1 is a criterion for the essential infeasibility of (D).

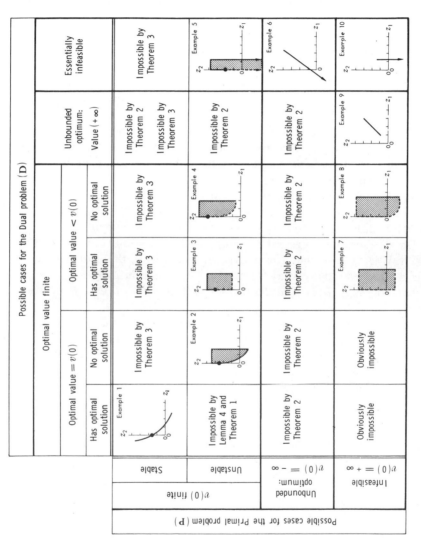

DIAG. 1

THEOREM 4. (D) *is essentially infeasible if and only if* $v(y) = -\infty$ *for some y.*

Proof. We shall prove the contrapositive. Suppose that (D) is essentially feasible, so that there exists a vector $\bar{u} \geq 0$ and a scalar M such that $f(x) + \bar{u}^t g(x) \geq M$ for all $x \in X$. Let y be an arbitrary point in Y. Then

$$f(x) \geq f(x) + \bar{u}^t(g(x) - y) \geq M - \bar{u}^t y$$

for all $x \in X$ such that $g(x) \leq y$. Taking the infimum of $f(x)$ over the indicated values of x, we obtain $v(y) \geq M - \bar{u}^t y > -\infty$. This proves that (D) is essentially feasible only if $v(y) > -\infty$ for all $y \in Y$.

Suppose now that $v(y)$ is not $-\infty$ anywhere on Y. To show that (D) is essentially feasible, it is enough to show that v has a subgradient $\bar{y}$ at some point $\bar{y}$ in Y, for then, by reasoning as in the first part of the proof of Lemma 4, we may demonstrate that $-\bar{y}$ is essentially feasible in (D). Let y be any point in Y, and put $\bar{y} = y + 1$, where 1 is an m-vector with each component equal to unity. Obviously, $\bar{y}$ is in the interior of Y, and $v(\bar{y})$ is finite by supposition. Since v is convex, therefore, by a known property of convex functions (see the remark following Lemma 1) it must have a subgradient at $\bar{y}$. Thus (D) is essentially feasible if $v(y) > -\infty$ for all $y \in Y$. The proof is now complete.

Since the convexity of v implies that it has value $-\infty$ at *some* point in Y if and only if it has value $-\infty$ at *every interior* point in Y, we see that (D) will be essentially infeasible if and only if $v(y) = -\infty$ on the whole interior of Y.

The next result is suggested by the apparent impossibility of altering Examples 7 and 8 of Diag. 1 so that I is strictly separated from the ordinate (i.e., so that $y = 0$ is bounded strictly away from Y; cf. [20, Theorem 2(b)]).

THEOREM 5. *If* (P) *is infeasible and the optimal value of* (D) *is finite, then* 0 *is in the closure of Y.*

Proof. Suppose, contrary to what we wish to show, that $y = 0$ is not in the closure of Y. Then, by the convexity of Y, 0 can be strictly separated from it by a hyperplane with normal p, say: $p^t y \geq \varepsilon > 0$ for all $y \in Y$. Certainly $p \geq 0$, for otherwise some component of y could be taken sufficiently large to violate $p^t y > 0$. Since $g(x) \in Y$ for all $x \in X$, we obtain

$$\underset{x \in X}{\text{infimum}} \{p^t g(x)\} > 0.$$

Let $u \geq 0$ be any vector that is essentially feasible in (D), so that

$$\underset{x \in X}{\text{infimum}} \{f(x) + u^t g(x)\} > -\infty.$$

Then $u + \theta p$ is also essentially feasible in (D) for any scalar $\theta \geq 0$, and

$$\underset{x \in X}{\text{infimum}} \{f(x) + (u + \theta p)^t g(x)\} \geq \underset{x \in X}{\text{infimum}} \{f(x) + u^t g(x)\} + \theta \underset{x \in X}{\text{infimum}} \{p^t g(x)\}.$$

By letting $\theta \to \infty$, we obtain the contradiction that the optimal value of (D) is $+\infty$. Hence 0 must be in the closure of Y, and the proof is complete.

This shows that the cases represented by Examples 7 and 8 (namely, (P) infeasible and a finite optimal value for (D)) can occur only if some arbitrarily small perturbation of y away from 0 will restore (P) to feasibility. Of course, if Y is closed (e.g., X compact and g continuous), then these cases cannot occur.

A similar result is suggested by the conspicuous feature which Examples 2–5 of Diag. 1 have in common : the ordinate touches I but never passes through its interior. In other words, $y = 0$ is always a boundary point of Y. This turns out to be true in general when (P) has a finite optimal value but is unstable, as follows from the next theorem.

THEOREM 6. *If $v(0)$ is finite and $y = 0$ is an interior point of Y, then (P) is stable.*

Proof. Since v is convex and 0 is in the interior of Y, it must have a subgradient at this point (see the remark following Lemma 1). Apply Lemma 2.

The converse of this theorem is *not* true ; that is, the stability of (P) does not imply that 0 is an interior point of Y. A counterexample is provided by Example 1 with the portion of I to the left of the dot removed.

The condition that 0 is in the interior of Y can be thought of as a "constraint qualification" which implies stability (provided $v(0)$ is finite). It is equivalent to the classical qualification introduced by Slater that there exist a point $x^0 \in X$ such that $g_i(x^0) < 0$, $i = 1, \cdots, m$. In this event, it follows from Theorem 6 that only the cases represented by Examples 1 and 6 of Diag. 1 can obtain.

5.2. Duality gaps and the continuity of v. There is one more result suggested by the examples of Diag. 1 that we wish to discuss. It pertains to the possibility of a duality gap, or difference in the optimal values of (P) and (D). From Diag. 1 we can immediately observe that this can happen when (P) is feasible only if (P) is unstable, although it need not happen if (P) is unstable. Geometric considerations suggest that any difference in optimal values must be due to a lack of continuity of v at $y = 0$ (a convex function can be discontinuous at points on the boundary of its domain). And indeed this important result is essentially so, as we now show.

THEOREM 7 (Continuity). *Let $v(0)$ be finite. Then the optimal values of (P) and (D) are equal if and only if v is lower semicontinuous at $y = 0$.*

Proof. The first part of the proof makes use of the function w,

$$w(y) \triangleq \underset{u \geq 0}{\text{supremum}} \left[\underset{x \in X}{\text{infimum}} \ f(x) + u^t(g(x) - y) \right],$$

which is to be interpreted as the optimal value of the dual problem of (P) modified to have right-hand side y rather than 0. Certainly w is a convex function on R^m, for it is the pointwise supremum of a collection of functions that are linear in y.

Suppose that v is lower semicontinuous at $y = 0$. It follows that v is finite on the interior of Y. Let $\langle y^v \rangle$ be the following sequence in the interior of Y converging to $0 : y^v = 1/v$, where 1 is an m-vector with each component equal to unity. It follows from Theorem 6 that (P), modified to have right-hand side y^v instead of 0, must be stable. By part (b) of Theorem 3, therefore, we conclude that $v(y^v) = w(y^v)$ for all v. We then have

$$w(0) \geq \underset{v \to \infty}{\lim \inf} \ w(y^v) = \underset{v \to \infty}{\lim \inf} \ v(y^v) \geq v(0) \geq w(0),$$

where the first inequality follows from the convexity of w, the second from the lower semicontinuity of v, and the last from the weak duality theorem. Hence $v(0)$ must equal $w(0)$, the optimal value of (D), and the "only if" part of the theorem is proved.

Now assume that $v(0)$ equals the optimal value of (D). We must show that v is lower semicontinuous at $y = 0$. Suppose to the contrary that there exists a sequence $\langle y^v \rangle$ of points in Y converging to 0 such that $\langle v(y^v) \rangle \to \bar{v} < v(0)$. We may derive the contradiction $\bar{v} \geq$ optimal value of (D) as follows. Since $f(x) + u^t g(x) \leq f(x) + u^t y$ holds for all $u \geq 0$, $y \in Y$ and $x \in X$ such that $g(x) \leq y$, we may take the infimum of both sides to obtain

$$\underset{x \in X}{\text{infimum}} \{ f(x) + u^t g(x) | g(x) \leq y \} \leq v(y) + u^t y \quad \text{for all } u \geq 0 \quad \text{and} \quad y \in Y.$$

It follows that

$$\left[\underset{x \in X}{\text{infimum}} \, f(x) + u^t g(x) \right] \leq v(y) + u^t y \quad \text{for all } u \geq 0 \quad \text{and } y \in Y.$$

Hence,

$$\left[\underset{x \in X}{\text{infimum}} \, f(x) + u^t g(x) \right] \leq \lim_{v \to \infty} (v(y^v) + u^t y^v) = \bar{v} \quad \text{for all } u \geq 0.$$

Taking the supremum over $u \geq 0$, we obtain the desired contradiction: optimal value of (D) $\leq \bar{v}$. This completes the proof.

Theorem 7 motivates the need for conditions which imply the lower semicontinuity of v at 0. The following result is of fundamental interest in this regard.

THEOREM 8. *Assume that X is closed, f and g are continuous on X, the optimal value of* (P) *is finite, and* $\{ x \in X : g(x) \leq 0 \text{ and } f(x) \leq \alpha \}$ *is bounded and nonempty for some scalar $\alpha \geq v(0)$. Then v is lower semicontinuous at $y = 0$.*

The boundedness hypothesis is obviously satisfied if, as is frequently the case in applications, the feasible region of (P) is bounded (put $\alpha = v(0) + 1$). If the boundedness of the feasible region is in question but (P) has an optimal solution, then the hypothesis holds if the optimal solution is unique or, more generally, if the set of alternative optimal solutions is bounded (put $\alpha = v(0)$). If the boundedness of the feasible region and the existence of an optimal solution of (P) are in question, then the hypothesis holds if a set of alternative ε-optimal solutions is bounded (put $\alpha = v(0) + \varepsilon$).

The proof of Theorem 8 depends on the following fundamental property of convex functions (cf. [12, p. 93]).

LEMMA 6. *Let $\phi(\cdot)$ be a convex continuous real-valued function on a closed convex set $X \subseteq E^n$. Define $\Phi_\varepsilon \triangleq \{ x \in X : \phi(x) \leq \varepsilon \}$, where ε is a scalar. If Φ_ε is bounded and nonempty for $\varepsilon = 0$, then it is bounded for all $\varepsilon > 0$.*

Proof. Let $\varepsilon > 0$ be fixed arbitrarily, and let x_0 be any point in Φ_0. The assumptions imply that Φ_ε is closed and convex, and that there exists a scalar $M > 0$ such that $\| x - x_0 \| < M$ for all $x \in \Phi_0$. Suppose, contrary to the desired conclusion, that there is an unbounded sequence $\langle x^v \rangle$ of points in Φ_ε. The direction vectors $(x^v - x_0)/\| x^v - x_0 \|$ must converge subsequentially to a limit, say Δ. Furthermore, $(x_0 + \theta \Delta) \in \Phi_\varepsilon$ for all $\theta \geq 0$, since Φ_ε is closed and contains a sequence of points whose limit is $x_0 + \theta \Delta$ (such a sequence is $\langle x_0 + \theta(x^v - x_0)/\| x^v - x_0 \| \rangle$ for v such that $\theta/\| x^v - x_0 \| \leq 1$—remember that Φ_ε is convex). Define $\lambda = \phi(x_0 + M\Delta)/2\varepsilon$. Clearly $0 < \lambda < 1$, and $(x_0 + M\Delta) = \lambda(x_0 + (M/\lambda)\Delta) + (1 - \lambda)x_0$.

The convexity of ϕ therefore implies

$$\phi(x_0 + M\Delta) \leq \lambda\phi[x_0 + (M/\lambda)\Delta] + (1 - \lambda)\phi(x_0).$$

Rearrangement of terms yields the key inequality

$$\phi\left[x_0 + \frac{M}{\lambda}\Delta\right] \geq \frac{1}{\lambda}\phi(x_0 + M\Delta) - \frac{(1 - \lambda)}{\lambda}\phi(x_0).$$

Since the choice of λ implies that the first term on the right has value 2ε, and since the second term is obviously nonnegative, this inequality directly contradicts the known fact that $[x_0 + (M/\lambda)\Delta] \in \Phi_\varepsilon$. Hence our supposition that Φ_ε is unbounded must be erroneous, and the lemma is proved.

Proof of Theorem 8. Suppose, contrary to the conclusion, that there exists a sequence $\langle y^v \rangle$ of points in Y such that $\langle y^v \rangle \to 0$ and $\langle v(y^v) \rangle \to \bar{v} < v(0)$. For each v, there exists a point $x^v \in X$ such that $g(x^v) \leq y^v$ and $v(y^v) \leq f(x^v) \leq v(y^v) + (1/v)$ (the infimal value $v(y^v)$ can be approached as closely as desired). Clearly $\langle f(x^v) \rangle \to \bar{v}$, and our assumptions imply that, for all v sufficiently large x^v must be in the set

$$\Xi \triangleq \{x \in X : g_i(x) \leq \varepsilon, i = 1, \cdots, m \text{ and } f(x) \leq v(0)\}$$

for any fixed $\varepsilon > 0$. But repeated application of Lemma 6 reveals that Ξ is bounded, and so $\langle x^v \rangle$ must be a bounded sequence. We may therefore assume (taking a subsequence if necessary) that $\langle x^v \rangle$ converges, say to $\bar{x}$. By the closedness of X and the continuity of g and f, we have $\bar{x} \in X$, $g(\bar{x}) \leq 0$, and $f(\bar{x}) = \lim f(x^v) = \bar{v}$. Thus $\bar{x}$ is feasible in (P) and $\bar{v} = f(\bar{x}) \geq v(0)$. But this contradicts the supposition $\bar{v} < v(0)$, and so v must be lower semicontinuous at $y = 0$.

5.3. A converse duality theorem. Lemma 6 is also the key to the following important partial converse to the strong duality theorem.

THEOREM 9 (Converse duality). *Assume that X is closed and that f and g are continuous on X. If* (D) *has an optimal solution* u^*, $f + (u^*)^t g$ *has a unique minimizer* x^* *over X, and x^* is feasible in* (P), *then x^* is the unique optimal solution of* (P), u^* *is an optimal multiplier vector, and* (P) *is stable.*

Proof. Since (D) has an optimal solution, we see from Diag. 1 that only the cases represented by Examples 1, 3 and 7 are possible. The last case is precluded by the assumption that x^* is feasible in (P). If we can show that v is lower semicontinuous at $y = 0$, then by Theorem 7 the first case must obtain and (P) must be stable. If we can also show that (P) has an optimal solution, then part (d) of Theorem 3 will imply that x^* must be the unique optimal solution, for it is the unique minimizer of $f + (u^*)^t g$ over X. Part (c) of Theorem 3 and Theorem 1 will also imply that u^* is an optimal multiplier vector for (P).

Thus our task is to demonstrate that v is lower semicontinuous at $y = 0$ and that (P) admits an optimal solution. To accomplish this, apply Lemma 6 with $\phi(x)$ equal to $f(x) + (u^*)^t g(x) - f(x^*) - (u^*)^t g(x^*)$. It follows that the set

$$\Phi_\varepsilon \triangleq \{x \in X : f(x) + (u^*)^t g(x) \leq f(x^*) + (u^*)^t g(x^*) + \varepsilon\}$$

is bounded for all $\varepsilon > 0$ (Φ_0 is identical with x^*).

To see that (P) has an optimal solution, let $\langle x^v \rangle$ be a feasible sequence such that $\langle f(x^v) \rangle \to v(0)$, and put $\varepsilon = v(0) + 1 -$ optimal value (D). It is easily verified that $x^v \in \Phi_\varepsilon$ for all v sufficiently large, and consequently the problem:

$$\underset{x \in \Phi_\varepsilon}{\text{Minimize}} \quad f(x) \quad \text{subject to} \quad g(x) \leq 0,$$

whose feasible region is contained in that of (P), must have the same optimal value. Since this optimal value is actually achieved (the minimand is continuous and the feasible region is compact), this demonstrates the existence of an optimal solution to (P). The lower semicontinuity of v at $y = 0$ follows from an easy modification of Theorem 8 in which the proof uses Φ_ε (with the same choice of ε) in place of Ξ. This completes the proof.

If $f + (u^*)'g$ is strictly convex in some neighborhood of x^*, it must have a unique minimizer over X; but the reverse implication need not hold. Of course, one would expect $f + (u^*)'g$ to be strictly convex near x^* when (P) is a nonlinear program, as only one of the functions f and g_i such that $u_i^* > 0$ need be strictly convex near x^* for this to be so. Another way of rationalizing the uniqueness of the minimizer of $f + (u^*)'g$ over X when (P) involves nonlinear functions is to apply Theorem 10 of § 7 with $\varepsilon = 0$; it then follows that f and each g_i with a positive multiplier would have to be *linear* over the set of alternative minimizers.

A possibly useful observation on Theorem 9 is that the uniqueness hypothesis on x^* can be weakened somewhat with the help of the concept of *g-uniqueness*.[3] If the set of all points x with some particular property is nonempty and g is constant on this set, then this set is said to be *g-unique*. It is *g*-uniqueness of x^*, rather than uniqueness, which is essential in Theorem 9. If in the hypotheses we substitute "and the set X^* of all minimizers of $f(x) + (u^*)'g(x)$ over X is bounded and *g*-unique," then the conclusion still holds with the substitution "then X^* is the set of optimal solutions of (P)."

It is interesting to note how the conclusions of Theorem 9 change if x^* is not *g*-unique but the set X^* of all minimizers of $f + (u^*)'g$ over X is still bounded. The set Φ_ε used in the proof remains bounded by Lemma 6, since $\Phi_0 \equiv X^*$ is nonempty and bounded. It follows that (P) is stable and has an optimal solution, u^* is an optimal multiplier vector, and the optimal solution set of (P) coincides with the points of X^* which also satisfy $g(x) \leq 0$ and $(u^*)'g(x) = 0$.

If X is bounded, then the hypothesis "x^* is feasible in (P)" can be omitted because its only role in the proof of Theorem 9—to ensure that (P) is feasible— can be played by Theorem 5 (Y must now be closed).

6. Relations to previous duality results. In this section we examine in more detail the relationships between our results and previous work on duality theory in linear, quadratic, and nonlinear programming. Rather than attempting an exhaustive survey or even citation of the literature, which by now is quite extensive,[4] we select several key papers as representatives for comparison. These are the well-known papers by Dorn [8], Wolfe [34], Stoer [28], Mangasarian and Ponstein [25], and Rockafellar [26]. The first paper is representative of the results that can be obtained for the special case of quadratic programming; the second of

---

[3] The usefulness of this concept is brought out more clearly in § 7.3.

[4] See the extensive bibliographies of [24, Chap. 8] and [26]. For a dual problem ostensibly quite different from the one considered here, see also Charnes, Cooper and Kortanek [4] (their "Farkas–Minkowski property" appears to imply stability when $v(0)$ is finite).

the results that can be obtained by means of the differential calculus and the classical results of Kuhn and Tucker; the third and fourth of the results obtainable by applying general minimax theorems; and the fifth of the more recent results obtainable by applying the theory of conjugate convex functions.

6.1. Linear programming. Let the primal linear program be:

$$\text{Minimize}_{x \geq 0} \quad c^t x \quad \text{subject to} \quad Ax \geq b.$$

With the identifications $f(x) \equiv c^t x$, $g(x) \equiv b - Ax$ and $X \equiv \{x \in R^n : x \geq 0\}$, (D) becomes:

$$\text{Maximize}_{u \geq 0} \quad [\text{infimum}_{x \geq 0} \; c^t x + u^t(b - Ax)].$$

Observe that the maximand has value $u^t b$ if $(c^t - u^t A) \geq 0$, and $-\infty$ otherwise. That is, $(c^t - u^t A) \geq 0$ is a necessary and sufficient condition for essential feasibility, so that (D) may be rewritten

$$\text{Maximize}_{u \geq 0} \quad u^t b \quad \text{subject to} \quad u^t A \leq c^t.$$

This, of course, is the usual dual linear program.

The stability of the primal problem when it has finite optimal value is a consequence of the fact that its constraints are all linear (the perturbation function v is piecewise-linear with a finite number of "pieces"). Thus Theorems 1 and 3 apply.

The duality theorem and the usual related results of linear programming are among those now at hand, either as direct specializations or easy corollaries of the results given in previous sections. It is perhaps surprising that, even in the heavily trod domain of linear programming, the geometrical interpretation of the dual given in § 4 does not seem to be widely known.

It is interesting to examine what happens if one dualizes with respect to only a subset of the general linear constraints. Suppose, for example, that the general constraints $Ax \geq b$ are divided into two groups, $A_1 x \geq b_1$ and $A_2 x \geq b_2$, and that we dualize only with respect to the second group; i.e., we make the identifications $f(x) \equiv c^t x$, $g(x) \equiv b_2 - A_2 x$ and $X \equiv \{x \in R^n : x \geq 0 \text{ and } A_1 x \geq b_1\}$. Then the new primal problem is still stable, and the dual problem becomes:

$$\text{Maximize}_{u_2 \geq 0} \left[\text{infimum}_{\substack{x \geq 0 \\ A_1 x \geq b_1}} c^t x + u_2^t(b_2 - A_2 x) \right].$$

Formidable as it looks, this problem is amenable to solution by at least three approaches, all of which can be effective when applied to specially structured problems. In terminology suggested by the author in [17], the first approach is via the piecewise strategy, of which Rosen's "primal partition programming" scheme may be regarded an example [17, § 4.2]; the second approach is via outer linearization of the maximand followed by relaxation (Dantzig–Wolfe decomposition may be regarded as an example [17, § 4.3]); the third approach is via a feasible directions strategy [19]. The point to remember is that the dual of a linear program need not be taken with respect to *all* constraints, and that judicious selection in this regard allows the exploitation of special structure. This point is probably even more important in the context of structured nonlinear programs. It has been stressed previously by Falk [10] and Takahashi [29].

6.2. Quadratic programming: Dorn [8]. Let the primal quadratic program be:

$$\text{Minimize}_{x} \quad \tfrac{1}{2}x'Cx - c'x \quad \text{subject to } Ax \leqq b,$$

where C is symmetric and positive semidefinite. The dual (D) with respect to all constraints is

$$\text{Maximize}_{u \geqq 0} \quad [\text{infimum}_{x} \, q(x;u)],$$

where

$$q(x;u) \triangleq \tfrac{1}{2}x'Cx - u'b + (u'A - c')x.$$

In what follows, we shall twice invoke the fact [14, p. 108] that a quadratic function achieves its minimum on any closed polyhedral convex set on which it is bounded below.

Consider the maximand of (D) for fixed u. The function $q(\cdot \, ; u)$, being quadratic, is bounded below if and only if its infimum is achieved, which in turn can be true if and only if the gradient of q with respect to x vanishes at some point. Thus $\inf_x q(x;u)$ equals $-\infty$ if there is no x satisfying $\nabla_x q(x;u) \equiv x'C + (u'A - c') = 0$; otherwise, it equals $\tfrac{1}{2}x'Cx - u'b + (-x'C)x$ for any such x (an obvious algebraic substitution has been made). We can now rewrite the dual problem as:

$$\text{Maximize}_{u \geqq 0} \quad [-u'b - \tfrac{1}{2}x'Cx \quad \text{for any } x \text{ satisfying } x'C + (u'A - c') = 0]$$

subject to

$$x'C + (u'A - c') = 0 \quad \text{for some } x.$$

But this is equivalent in the obvious sense to:

$$\text{Maximize}_{\substack{u \geqq 0 \\ x}} \quad -u'b - \tfrac{1}{2}x'Cx \quad \text{subject to } x'C + u'A - c' = 0.$$

In this way do the primal variables find their way back into the dual. This is precisely Dorn's dual problem.

The linearity of the constraints of the primal guarantees stability whenever the optimal value is finite, and so Theorems 1 and 3 apply. Dorn's dual theorem asserts that if x^* is optimal in the primal, then there exists u^* such that (u^*, x^*) is optimal in the dual and the two extremal values are equal. This is a direct consequence of Theorem 3. Dorn's converse duality theorem asserts that if $(\hat{u}, \hat{x})$ are optimal in the dual, then some $\bar{x}$ satisfying $C(\bar{x} - \hat{x}) = 0$ is optimal in the primal and the two extremal values are equal. To recover this result, we first note by Theorem 2 that the primal minimand must be bounded below, and hence there must be a primal optimal solution, say $\bar{x}$. By Theorem 3, the extremal values are equal, and $\bar{x}$ must be an unconstrained minimum of $q(x \, ; \hat{u})$, i.e., $\bar{x}'C + \hat{u}'A - c' = 0$. But $\hat{x}'C + \hat{u}'A - c' = 0$ also holds, and so $\bar{x}'C = \hat{x}'C$.

mmetric dual of the special quadratic
nain results.

ing: **Wolfe [34].** The earliest and
lts for differentiable convex programs
all functions to be convex and *differ-*
ual for (P):

$$+ u^t g(x)$$

$$_i(x) = 0.$$

$-$*Tucker constraint qualification.* The
asserts that if x^0 is optimal in (P),
al in (W) and the extremal values are
$_i$ to be linear, asserts that the optimal
nfeasible.

rs, we must examine the relationship
$\bar{u} \geq 0$, $(\bar{u}, \bar{x})$ is feasible in (W) if and
$+ \bar{u}^t g$. In terms of the dual variables

$$f(x) + u^t g(x)]$$

minimum of $f + u^t g$ is achieved for
some x.
strong sense: $(\bar{u}, \bar{x})$ is feasible in (W)
objective function values are equal;
point $\bar{x}$ such that $(\bar{u}, \bar{x})$ is feasible in
lues are equal. Problem (W.1) is, of
constraint on u. We are now in a
theorem follows immediately from
straint on u in (W.1) can only depress
). Wolfe's second theorem follows
he strong duality theorem, since the
s stability. His third theorem would
t not for the extra constraint on u in
statement of the theorem. The extra
line of reasoning, and although one
results of this paper, it would not be
warrant presentation here.
fficult to deal with computationally;
its constraint set involves nonlinear
vex. Another difficulty is that (W) is
duality gap, since it has (as revealed
by (W.1)) an extra constraint on u.

6.4. Stoer [28] and Mangasarian and Ponstein [25]. The natural generalization of Wolfe's dual when some functions in (P) are not differentiable or the set X is not all of R^n is the following, which might be called the general Wolfe dual:

(GW)   Maximize $f(x) + u^t g(x)$
                              $\substack{u \geq 0 \\ x}$

subject to:

$$x \text{ minimizes } f + u^t g \text{ over } X.$$

This problem bears the same relation to (D) as (W) does; the more general version of the intermediary program (W.1) that is appropriate here should be evident. Although (GW) is generally inferior to (D) for much the same reason that (W) is, it is nevertheless worthwhile to review some of the work that has been addressed to this dual.

The landmark paper treating (GW) is by Stoer [28], whose principal tool is a general minimax theorem of Kakutani. The possibility of using minimax theorems in this connection is due, of course, to the existence of an equivalent characterization of the optimality conditions for (P) as a constrained Lagrangean saddle point. Stoer's results are shown to generalize many of those obtained via the differential calculus by numerous authors in the tradition of Wolfe's paper. Because of certain technical difficulties inherent in his development, however, we shall examine Stoer's results as reworked and elaborated upon by Mangasarian and Ponstein [25]. To bring out the essential contributions of this work, we shall take considerable license to paraphrase.

Aside from some easy preliminary results which do not depend on convexity— namely, a weak duality theorem and an alternative characterization of a constrained Lagrangean saddle point (see the discussion following our Definition 3)— the Stoer–Mangasarian–Ponstein results relating (P) and (GW) can be paraphrased as three theorems, all of which require f and g to be convex and *continuous*, and X to be convex and *closed*. The first [25, Theorem 4.4a] is: Assuming (P) has an optimal solution $\bar{x}$, an optimal multiplier vector $\bar{u}$ exists if and only if $f(x) + u^t g(x)$ has the so-called "low-value property" at $(\bar{x}, \bar{u})$. We shall not quote in detail this rather technical property, but we do observe that, in view of our Theorem 1, the low-value property must be entirely equivalent (when all functions are continuous and X is closed) to the condition that (P) is stable.

The second theorem [25, Theorem 4.4b] is: Assuming (GW) has an optimal solution $(\bar{x}, \bar{u})$, there exists a minimizer x^0 of $f + \bar{u}^t g$ over X satisfying $\bar{u}^t g(x^0) = 0$ such that x^0 is optimal in (P) if and only if $f(x) + u^t g(x)$ has the so-called "high-value property" at $(\bar{x}, \bar{u})$. The high-value property is also quite technical, but its significance can be brought out by comparing the theorem with the following immediate consequence of Theorems 1 and 3 of this study: There exists an optimal solution of (D) and for any optimal $\bar{u}$ there is a minimizer x^0 of $f + \bar{u}^t g$ over X satisfying $\bar{u}^t g(x^0) = 0$ such that x^0 is optimal in (P), if and only if (P) is stable and has an optimal solution. It follows that the high-value property holds and (GW) has an optimal solution if and only if (P) is stable and has an optimal solution (when f and g are continuous and X is closed). The demonstration is straightforward, and makes use of the evident fact that if $\bar{u}$ is optimal in (D) and $\bar{x}$ minimizes $f + \bar{u}^t g$ over X, then $(\bar{x}, \bar{u})$ must be optimal in (GW).

The third result is a strict converse duality theorem, a counterpart of our Theorem 9: If $(\bar{x}, \bar{u})$ is an optimal solution of (GW) and $f(x) + \bar{u}^t g(x)$ is strictly convex in some neighborhood of $\bar{x}$, then $\bar{x}$ is an optimal solution of (P) and $(\bar{x}, \bar{u})$ satisfies the optimality conditions for (P). The difference between this and Theorem 9, besides the fact that it addresses (GW) rather than (D), is the slightly stronger hypothesis that $f + \bar{u}^t g$ be strictly convex near $\bar{x}$ (rather than simply requiring the minimizer of $f + \bar{u}^t g$ over X to be unique or just g-unique).

This discussion casts suspicion on the need for general minimax theorems as a means of obtaining strong results in duality. Such an approach may even be inadvisable, as it seems to lean toward (GW) rather than (D), and toward technical conditions less convenient than stability.

6.5. Rockafellar [26]. Finally we come to the outstanding work of Rockafellar, whose methods rely heavily upon Fenchel's theory of conjugate convex functions [11]. To make full use of this theory, it is assumed in effect that f and g_i are *lower semicontinuous* on X (so that the convex bifunction associated with (P) will be closed as well as convex, as assumed in Rockafellar's development).[5] The theory of conjugacy then yields the fundamental relationships between (P) and (D). By way of comparison, we note that one can readily deduce Lemmas 1 through 5 and Theorems 1 through 7 of this study from his results. This deduction utilizes the equivalence of Definitions 6 and 6' for the concept of stability, and the equivalence between the optimality conditions of (P) and a constrained saddle point of the Lagrangean. The content of our Theorems 8 and 9 is not obtained, but Rockafellar does give some additional results not readily obtainable by our methods. Namely, the dual of (P) in a certain conjugacy sense is again (P); the optimal solutions of (P) are the subgradients of a certain perturbation function associated with (D); and v is lower semicontinuous at 0 if and only if the perturbation function associated with (D) is upper semicontinuous at the point of null perturbation. The significance of these additional results for applications is not clear, although they are certainly very satisfying in terms of mathematical symmetry.

Although (D) is the natural dual problem of (P) corresponding to the perturbation function v, it is interesting to note that other perturbation functions give rise to other duals to which Rockafellar's results apply immediately. It seems likely that the methods of this paper could be adapted to deal with other perturbation functions too, but as yet this has not been attempted.

7. Computational applications. This section studies a number of issues that arise if one wishes to obtain a numerical solution of a convex program via its dual—that is, if one is interested in "dual" methods for solving (P). By a dual method we mean one that generates a sequence of essentially feasible solutions that converges in value to the optimal value of (D). Such a sequence yields, by the weak duality theorem, an improving sequence of lower bounds on the optimal value of (P).

---

[5] Rockafellar has pointed out to the author in a private communication (October 13, 1969) that bifunction closedness, and hence the semicontinuity assumption on f and g_i, can be dropped except for the "additional" results referred to below. He also pointed out that results closely related to our Theorems 8 and 9 appear in his forthcoming book *Convex Analysis* (Princeton University Press).

The possible pitfalls encountered by dual methods include these: The dual may fail to be essentially feasible, even though (P) has an optimal solution; or it may be essentially feasible but fail to have an optimal solution; or it may have an optimal solution but its optimal value may be less than that of (P) (this can invalidate the obvious natural termination criterion when an upper bound on the optimal value of (P) is at hand). None of these pitfalls can occur if (P) is stable, thanks to the strong duality theorem. For this reason, and also because this property usually holds anyway, we shall *assume for the remainder of the section that* (P) *is stable.* We shall also assume that (P) has an optimal solution.

7.1. Methods for solving (D). There are several ways one may go about solving (D). Some of these are appropriate only when (P) has quite special structure, as when all functions are linear or quadratic. Others can be employed under quite general assumptions. It is perhaps fair to say that most of these methods fall into two major categories: methods of feasible directions and methods of tangential approximation. Both are based on the easily verified fact that, for any $\bar{u} \geq 0$, if $\bar{x}$ achieves the infimum required to evaluate the maximand of (D) at $\bar{u}$ then $g(\bar{x})$ is a subgradient of the maximand at $\bar{u}$; that is,

$$\operatorname*{infimum}_{x \in X} \{ f(x) + u'g(x) \} \leq [f(\bar{x}) + \bar{u}'g(\bar{x})] + g'(\bar{x})(u - \bar{u}), \qquad \text{all } u \geq 0.$$

The feasible directions methods typically use $g(\bar{x})$ as though it were a gradient to determine a direction in which to take a "step." Instances of such methods are found in [3], [10], [19], [22], [29], [33]. The term "Lagrangean decomposition" is sometimes applied when the maximand of (D) separates (decomposes) into the sum of several independent infima of Lagrangean functions. The ancestor of this class of algorithms is an often-overlooked contribution by Uzawa [30]. The other principal class of methods for optimizing (D) is global rather than local in nature, as it uses subgradients of the maximand of (D) to build up a tangential approximation to it. It is well known that the decomposition method of Dantzig and Wolfe for nonlinear programming [6, Chap. 24] can be viewed as such a method. See also [16, § 6] and the "global" approach in [29].

It is beyond the scope of this effort to delve into the details of methods for optimizing (D). Rather, we wish to focus on questions of common interest to almost any algorithm that may be proposed for (D) as a means of solving (P). Specifically, we shall consider the possibility of numerical error in optimizing (D) and in minimizing $f + u'g$ over X for a given u. It is important to investigate the robustness of the resulting approximate solutions to (P) when there is numerical error of this kind. Hopefully, by making the numerical error small enough one can achieve an arbitrarily accurate approximation to an optimal solution of (P). We shall see that this is often, but not always, the case.

7.2. Optimal solution of (D) known without error. The simplest case to consider is the one in which an optimal solution u^* of (D) is known without error. Then by the strong duality theorem we know that the optimal solutions of (P) coincide with the solutions in x of the system:[6]

(i) x minimizes $f + (u^*)'g$ over X,

(ii) $(u^*)'g(x) = 0$,

(iv) $g(x) \leq 0$.

If (i) is known to have a unique solution, as is very often the case when (P) is a nonlinear program (actually, g-uniqueness is enough—see the discussion following Theorem 9), then (ii) and (iv) will automatically hold at the solution. Any sequence of points converging to the solution of (i) also converges to an optimal solution of (P). Thus, there appear to be no particular numerical difficulties.

If, on the other hand, the solution set of (i) is not unique, then (ii) or (iv) or both may be relevant. The following result will be useful.

THEOREM 10. *Let $u \geq 0$ and $\varepsilon \geq 0$ be fixed. Let X_ε be the set of ε-optimal solutions of the problem of minimizing $f + u'g$ over X; that is, let X_ε be the (convex) set of all points $\bar{x}$ in X such that*

$$f(\bar{x}) + u'g(\bar{x}) \leq \left[\underset{x \in X}{\text{infimum}} \, f(x) + u'g(x) \right] + \varepsilon.$$

Then f comes within ε, and each g_i with a positive multiplier comes within (ε/u_i) of being linear over X_ε in the following sense: x^1, $x^2 \in X_\varepsilon$ and $0 \leq \lambda \leq 1$ implies (let $\bar{\lambda} = 1 - \lambda$)

$$\lambda f(x^1) + \bar{\lambda} f(x^2) - \varepsilon \leq f(\lambda x^1 + \bar{\lambda} x^2) \leq \lambda f(x^1) + \bar{\lambda} f(x^2),$$

$$\lambda g_i(x^1) + \bar{\lambda} g_i(x^2) - \frac{\varepsilon}{u_i} \leq g_i(\lambda x^1 + \bar{\lambda} x^2) \leq \lambda g_i(x^1) + \bar{\lambda} g_i(x^2), \qquad i: u_i > 0.$$

Proof. The right-hand inequalities hold, of course, by convexity. Suppose that the first left-hand inequality fails for some $u \geq 0$, $\varepsilon \geq 0$, $x^1 \in X_\varepsilon$, $x^2 \in X_\varepsilon$, $0 \leq \lambda \leq 1$. Then

$$f(\lambda x^1 + \bar{\lambda} x^2) < \lambda f(x^1) + \bar{\lambda} f(x^2) - \varepsilon$$

which, when added to the other right-hand inequalities multiplied by the respective values of u_i, yields

$$f(\lambda x^1 + \bar{\lambda} x^2) + \sum_{i=1}^{m} u_i g_i(\lambda x^1 + \bar{\lambda} x^2)$$

$$< \lambda \left[f(x^1) + \sum_{i=1}^{m} u_i g_i(x^1) \right] + \bar{\lambda} \left[f(x^2) + \sum_{i=1}^{m} u_i g_i(x^2) \right] - \varepsilon.$$

Since X is convex, $\lambda x^1 + \bar{\lambda} x^2$ is in X and so the left-hand side has value greater than or equal to

$$\underset{x \in X}{\text{infimum}} \, f(x) + u'g(x).$$

Using the fact that x^1 and x^2 are in X_ε, however, we obtain the contradiction that

---

[6] The rubrics (i), (ii) and (iv) are used in order to maintain correspondence with the optimality conditions as listed in Definition 3.

the right-hand side is less than or equal to this value. Hence our supposition must fail.

A similar argument shows that the other inequalities of the conclusion of the theorem must hold, completing the proof.

We must distinguish two further possibilities when (i) does not have a unique solution: either a solution of (i) can be found without error, or it cannot. Suppose that an optimal solution $\bar{x}$ of (i) can be found. Then (i) is equivalent to:

(ia) $x \in X,$

(ib) $f(x) + (u^*)^t g(x) \leqq f(\bar{x}) + (u^*)^t g(\bar{x}).$

Theorem 10 with $u = u^*$ and $\varepsilon = 0$ yields the very useful result that (ii) is a *linear* constraint so long as x satisfies (ia) and (ib) ($(u^*)^t g(x) \equiv \sum u_i^* g_i(x)$, where the sum is taken over the indices such that $u_i^* > 0$). Thus, any of a number of convex programming algorithms could be used with $\bar{x}$ as the starting point to find a feasible solution of (ia), (ib), (ii), and (iv) and thereby solve (P). Suppose, on the other hand, that only an ε-optimal solution $\bar{x}$ of (i) can be found. Then (ia) and (ib) are no longer equivalent to (i), and (ii) is no longer linear over the solution set of (ia) and (ib). However, this solution set contains the solution set of (i) (because $f(\bar{x}) + (u^*)^t g(\bar{x})$ is larger than it ought to be); and Theorem 10 implies that $(u^*)^t g(x)$ is within $\# \varepsilon$ of being linear over it, where $\#$ is the number of indices for which $u_i^* > 0$. Hence, any convex programming algorithm which solves (ia), (ib) and (iv) exactly but (ii) only to linear approximation will find a feasible solution of (P) that is within $(\# + 1)\varepsilon$ of being optimal in (P). Therefore, by taking ε sufficiently small one can find a solution of (P) that is as near optimal as desired.

In summary, we see that solving (P) once an optimal solution of (D) is known poses no special numerical difficulties.

7.3. Optimal solution of (D) not known exactly. Suppose that a particular algorithm addressed to (D) generates a sequence $\langle u^v \rangle$ converging to an optimal solution u^*. If the minimizers of $f + (u^*)^t g$ over X are g-unique, we can obtain a quite satisfactory result concerning the recovery of an optimal solution of (P). In the absence of this assumption, however, an example will be given to show that things can go awry.

THEOREM 11. *Assume that f and each g_i is continuous on X, X is compact, (D) has an optimal solution u^*, and the minimizers of $f + (u^*)^t g$ over X are g-unique. Let $\langle u^v \rangle$ be any nonnegative sequence converging to u^* and $\langle x^v \rangle$ any sequence composed, for each v, of a minimizer of $f + (u^v)^t g$ over X. Then $\langle x^v \rangle$ has at least one convergent subsequence, and every such subsequence converges to an optimal solution of (P).*

Proof. Since $\langle x^v \rangle$ is in X and X is compact, there must be at least one convergent subsequence. For simplicity of notation, redefine $\langle x^v \rangle$ to coincide with any such subsequence. Let $X(u)$ be the set of all minimizers of $f + u^t g$ over X. Under the given assumptions it is known (e.g., [7, p. 19]) that $X(u)$ is an upper semicontinuous set-valued function of u at u^*; that is, $\langle u^v \rangle \to u^*$, $x^v \in X(u^v)$, $\langle x^v \rangle \to \bar{x}$ implies $\bar{x} \in X(u^*)$. But $X(u^*)$ is bounded and g-unique, and so by Theorem 9 (see also the ensuing discussion) $\bar{x}$ must be an optimal solution of (P). The proof is complete.

The conclusion of Theorem 11 is very comforting; but in practice one must still decide when to truncate the infinite process. We shall assume in the ensuing discussion that the hypotheses of Theorem 11 hold, except where explicitly weakened.

One natural termination criterion is based upon the easily demonstrated fact that x^v must be an optimal solution of the following approximation to (P):

(P^v) $$\text{Minimize}_{x \in X} \quad f(x) \quad \text{subject to } g(x) \leqq g(x^v).$$

The continuity of g, and the fact that $\langle x^v \rangle$ converges subsequentially to an optimal solution x^* of (P), implies that the right-hand side of (P^v) converges subsequentially to $g(x^*)$ as $v \to \infty$; hence one may terminate when v reaches a value for which the right-hand side of (P^v) is "sufficiently near" to being $\leqq 0$. How near is "sufficiently" near depends upon how precisely the g constraints of (P) really must be satisfied. If a perturbation in the right-hand side of certain of the constraints cannot be tolerated, however small the change, then it may be advisable to insist that such constraints be incorporated into X. In other words, it may be advisable not to dualize with respect to such constraints in the first place.

So far we have assumed that a true minimizer x^v of $f + (u^v)^t g$ over X could be found for each v. While this may be a reasonable assumption when X is a convex polytope and f and g are linear or quadratic functions, it is desirable in the interest of generality to be able to cope with numerical inaccuracy. Results as satisfactory as those obtained above seem quite elusive unless we suppose that a minimizer of $f + (u^v)^t g$ over X, say x^v, can be approached as closely as desired and, indeed, is approached more and more closely as v increases. Specifically, let us suppose that a point $\bar{x}^v$ in X can be found satisfying $\|\bar{x}^v - x^v\| \leqq \varepsilon^v$, say, where $\varepsilon^v \geqq 0$ and $\langle \varepsilon^v \rangle \to 0$. It follows easily that, for each subsequence of $\langle x^v \rangle$ converging to x^*, the corresponding subsequence of $\langle \bar{x}^v \rangle$ also converges to x^*. Thus, Theorem 11 holds with $\langle \bar{x}^v \rangle$ in place of $\langle x^v \rangle$. Of course, $\bar{x}^v$ is not optimal in (P̄^{v}), which we define to be (P^v) with $\bar{x}^v$ in place of x^v in the right-hand side. What is true, however, is that $\bar{x}^v$ is within $\bar{\varepsilon}$ of being optimal in (P̄^{v}) if it comes within $\bar{\varepsilon}$ of minimizing $f + (u^v)^t g$ over X (see [9]). Since the right-hand side of (P̄^{v}) converges subsequentially to $g(x^*)$ as $v \to \infty$, we can use (P̄^{v}) in much the same way as (P^v) to determine when to terminate, except that the magnitude of $\bar{\varepsilon}$ must also be considered.

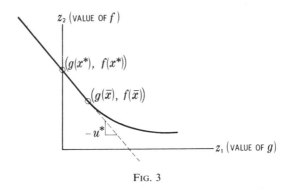

FIG. 3

Finally, we wish to emphasize the central role played by the assumption of Theorem 11 that the minimizers of $f + (u^*)^t g$ over X are g-*unique*. Failure of this assumption can lead to serious difficulties; see, for example, Fig. 3, in which any sequence $\langle u^v \rangle$ converging to u^* from below will lead to a sequence $\langle x^v \rangle$ with $g(\bar{x}) > 0$ for every subsequential limit point $\bar{x}$. Not only does $\bar{x}$ fail to solve (P), but $g(x^v) \geqq g(\bar{x}) > 0$ for all v, so that the natural termination criterion based on (P^v) will never be satisfied (unless of course the permissible tolerance in the right-hand side of (P) is sufficiently large).

8. Theoretical applications. Although the primary emphasis thus far has been on results of interest from the computational viewpoint, many of the results are also of interest from a purely theoretical point of view. Just as linear duality theory can be used to obtain many results in the theory of linear systems that do not appear to involve optimization at all (see, e.g., [6, § 6.4]), so does nonlinear duality theory readily yield many results in the theory of convex systems. In this section, we illustrate this fact by using our results to obtain new proofs of three known theorems. The first is a separation theorem for disjoint convex sets; the second is a characterization of a certain class of convex sets in terms of supporting half-spaces; and the third is a fundamental property of an inconsistent convex system. With only a modicum of ingenuity, by similar methods one may obtain many other interesting results, some of them perhaps new. Thus, nonlinear duality theory provides a unified approach to the derivation of a substantial body of theorems in the theory of convexity.

8.1. A separation theorem. Let X and $\tilde{X}$ be nonempty, closed, convex, disjoint subsets of R^k, and suppose further that X is bounded; then there exists a hyperplane in R^k that strictly separates X and $\tilde{X}$. A conventional proof of this theorem can be found, for example, in [2, p. 55], but we shall deduce it from Theorem 3. The other standard separation theorems can be obtained in a similar fashion. (The alert reader will recall that another separation theorem was used in the course of proving Theorem 3; what is being demonstrated in effect, then, is a kind of equivalence between duality theory and separation theory for convex sets.)

The hypotheses of the theorem certainly imply that the convex program

$$\text{Minimize}_{\substack{x \in X \\ \tilde{x} \in \tilde{X}}} \|x - \tilde{x}\|$$

has infimal value greater than 0. A convenient choice of norm is $\|x\| \triangleq \text{maximum} \{|x_1|, \cdots, |x_k|\}$. Then the above problem can be rewritten

$$\text{Minimize}_{\substack{x \in X \\ \tilde{x} \in \tilde{X} \\ \sigma}} \quad \sigma \quad \text{subject to} \quad \sigma \geqq x_i - \tilde{x}_i, \qquad i = 1, \cdots, k,$$
$$\sigma \geqq -x_i + \tilde{x}_i, \qquad i = 1, \cdots, k,$$

where σ is a scalar variable. Dualizing with respect to the (linear) constraints involving σ, we obtain from Theorem 3 that the dual problem

$$\text{Maximize}_{u \geqq 0} \left[\inf_{\substack{x \in X \\ \tilde{x} \in \tilde{X} \\ \sigma}} \sigma + \sum_{i=1}^{k} u_i(x_i - \tilde{x}_i - \sigma) + \sum_{i=1}^{k} u_{k+i}(-x_i + \tilde{x}_i - \sigma) \right]$$

has an optimal solution u^*, and that its optimal value is greater than 0. By taking advantage of the separability of the infimand with respect to x, $\tilde{x}$ and σ, we therefore have the key inequality

$$\left[\operatorname*{infimum}_{\sigma} \sigma\left(1 - \sum_{i=1}^{2k} u_i^*\right)\right] + \left[\operatorname*{infimum}_{x \in X} \sum_{i=1}^{k} (u_i^* - u_{k+i}^*)x_i\right]$$

$$+ \left[\operatorname*{infimum}_{\tilde{x} \in \tilde{X}} \sum_{i=1}^{k} - (u_i^* - u_{k+i}^*)\tilde{x}_i\right] > 0.$$

The infimum over σ must be 0 (i.e., $\sum_{i=1}^{2k} u_i^* = 1$ must hold), for otherwise it would be $-\infty$ and the inequality could not be true. Defining $\alpha_i = u_i^* - u_{k+i}^*$ and rearranging, the inequality then becomes

$$\operatorname*{infimum}_{x \in X} \sum_{i=1}^{k} \alpha_i x_i > \operatorname*{supremum}_{\tilde{x} \in \tilde{X}} \sum_{i=1}^{k} \alpha_i \tilde{x}_i.$$

This shows that X and $\tilde{X}$ are strictly separated by the hyperplane

$$\left\{ x \in R^k : \sum_{i=1}^{k} \alpha_i x_i = \alpha_0 \right\},$$

where α_0 is any scalar strictly between the values of the left- and right-hand sides of the rearranged inequality.

8.2. A characterization of the set Y. The set $Y \triangleq \{y \in R^m : g(x) \leqq y$ for some $x \in X\}$ has cropped up quite often in this study. Indeed, this kind of set arises frequently in mathematical programming when it is necessary to work with the collection of perturbations for which a perturbed problem has a feasible solution. It also arises when a set must be "projected" onto a subspace; in the special case above, Y can be thought of as being obtained by projecting the set $\{(x, y) \in R^{n+m} : g(x) - y \leqq 0, x \in X\}$ onto the R^m space associated with y (see [17, § 2.1]).

It is sometimes useful to be able to characterize such sets in terms of their supporting half-spaces. In [16, p. 24], a slightly weaker form of the following theorem is demonstrated by applying a result due to Bohnenblust, Karlin and Shapley (see [2, p. 64]). Assume that $g_1, \cdots, g_m$ are convex functions on the nonempty convex set $X \subseteq R^n$ and that Y is closed. Then $y \in Y$ if and only if y satisfies the system of linear constraints

$$\lambda^t y \geqq \operatorname*{infimum}_{x \in X} \lambda^t g(x), \qquad \text{all } \lambda \in \Lambda,$$

where $\Lambda \triangleq \{\lambda \in R^m : \lambda \geqq 0$ and $\sum_{i=1}^{m} \lambda_i = 1\}$. Furthermore, every constraint in this system describes a half-space that supports Y in the obvious sense.[7]

The only part of the conclusion that cannot be proved directly and easily is the assertion that $\bar{y} \in Y$ if $\bar{y}$ satisfies the given system of constraints. To prove this using nonlinear duality theory, we observe that if $\bar{y}$ satisfies the given system of

---

[7] That is, for each fixed $\bar{\lambda} \in \Lambda$, either there exists a point $\bar{y}$ in Y such that $\bar{\lambda}^t \bar{y} = \operatorname{infimum}_{x \in X} \bar{\lambda}^t g(x)$ or there exists a sequence $\langle y^\nu \rangle$ of points in Y such that $\lim_{\nu \to \infty} \bar{\lambda}^t y^\nu = \operatorname{infimum}_{x \in X} \bar{\lambda}^t g(x)$.

constraints, then (as the normalization of λ is immaterial) we have

$$\operatorname*{supremum}_{\lambda \geq 0} \left[\operatorname*{infimum}_{x \in X} \lambda'(g(x) - \bar{y}) \right] = 0.$$

This is easily recognized as the assertion that the dual of

$$\operatorname*{Minimize}_{x \in X} \quad 0'x \quad \text{subject to} \quad g(x) - \bar{y} \leq 0$$

has optimal value 0. Applying Theorem 5 to this "primal" problem yields, since $\{y : g(x) - \bar{y} \leq y \text{ for some } x \in X\}$ must be closed when Y is closed, that this problem must be feasible and hence that $\bar{y}$ must be in Y.

8.3. A fundamental property of an inconsistent convex system. Let $g_1, \cdots, g_m$ be convex functions on the nonempty convex set $X \subseteq R^n$. If the system $g_1(x) < 0$, $\cdots, g_m(x) < 0$ has no solution in X, then there exists an m-vector $u \geq 0$ such that $\sum_{i=1}^{m} u_i = 1$ and

$$\left[\operatorname*{infimum}_{x \in X} u'g(x) \right] \geq 0.$$

This is essentially the Fundamental Theorem on p. 62 of [2], where a proof relying on Helly's intersection theorem can be found. To deduce it from Theorem 3, we merely observe that the system of inequalities has no solution in X only if the convex program

$$\operatorname*{Minimize}_{\substack{x \in X \\ \sigma}} \quad \sigma \quad \text{subject to} \quad g_i(x) - \sigma \leq 0, \quad i = 1, \cdots m,$$

has infimal value ≥ 0. Since Slater's qualification is obviously satisfied, this program is stable and so by Theorem 3 we conclude that the dual program

$$\operatorname*{Maximize}_{u \geq 0} \left[\operatorname*{infimum}_{\substack{x \in X \\ \sigma}} \sigma + \sum_{i=1}^{m} u_i(g_i(x) - \sigma) \right]$$

has an optimal solution u^* with value ≥ 0. Hence

$$\left[\operatorname*{infimum}_{x \in X} (u^*)'g(x) \right] + \left[\operatorname*{infimum}_{\sigma} \sigma \left(1 - \sum_{i=1}^{m} u_i^* \right) \right] \geq 0,$$

where we have taken advantage of the separability of the infimum in the dual program. The second term must be 0—that is, $\sum_{i=1}^{m} u_i^* = 1$ must hold—for otherwise its value would be $-\infty$. Hence the first term is nonnegative and the theorem is proved.

A generalization of the Farkas–Minkowski theorem applicable to convex systems [2, p. 67] can be readily demonstrated by a very similar line of reasoning.

9. Opportunities for further research. It is hoped that what has been accomplished here will encourage further work in a similar vein. Much remains to be done.

In terms of importance, one could hardly do better than to work toward relaxing the convexity assumptions. We have pointed out several occasions on which these assumptions could be dispensed with entirely. For example, it is tantalizingly true that (D) is a concave program even without any assumptions at all on X, f and g. Furthermore, the astute reader may have noticed that the only role played by the assumed convexity of f, g and X in §§ 2 and 3 (and in a number of later results) is to guarantee via Lemma 2 that v and Y are convex. Thus the convexity assumptions on f, g and X can be weakened at least in Theorems 1 and 3, to the assumption that v and Y are convex. Perhaps a theory adequate for some purposes could be constructed under still weaker assumptions. Quite likely this can be done using the notions of pseudo- or quasi-convexity (e.g., [24, Chaps. 9, 10]) in place of convexity, but the real challenge would be to get along with considerably weaker assumptions. Published efforts along these lines so far leave a lot to be desired in terms of potential applicability, mainly because global results in the absence of global properties like convexity seem to require assuming global knowledge that is overly difficult to have in practice. Perhaps a local theory is all one can hope for without convexity-like assumptions.

Important opportunities are also to be found in studying questions such as those treated in § 7, relating to the robustness of computational methods addressed to the dual problem in the face of numerical error. For example, to what extent does an inability to minimize the Lagrangean function exactly for a given value of u disrupt convergence to an optimal solution of (D)? Probably such studies will have to be carried out in the context of the various specific dual methods that have been proposed. One successful study in this vein is Fox [13].

A natural extension of the theory developed here would be the construction of parallel theories for perturbation functions other than v, perhaps even for an entire class of them. Rockafellar's work strongly suggests that this is possible. Some of the alternative choices for v might prove quite useful in widening the scope of applications of duality theory.

Another direction of possible extension would be toward more general linear vector spaces. This would open up applications to optimal control, continuous programming, and other infinite-dimensional problems. In fact, a good deal has already been accomplished along these lines (e.g., Luenberger [23], Rockafellar [27], Van Slyke and Wets [31]), particularly with reference to generalizations of the results of §§ 2 and 3 of this paper. The very recent treatment by Luenberger adopts a viewpoint quite similar to the one taken here, and is especially recommended to the interested reader.

A number of questions concerning economic significance and interpretation yet remain to be explored. The paper by Gale [15], which includes a discussion of optimal multiplier vectors and the concept of stability, is a fine example of what can be done. See also Balinski and Baumol [1] and Williams [32].

Finally, we should mention that a variety of opportunities for application exist even without further extensions. Many more theoretical applications of the kind illustrated in § 8 are possible. Illustrative of the possible computational applications is a recent nonlinear generalization by this writer of Bender's partition programming method [18].

Acknowledgments. The author expresses his appreciation to B. Fox, R. Grinold, R. Marsten, R. T. Rockafellar, and C. Swoveland for constructive criticisms of the manuscript.

REFERENCES

[1] M. L. BALINSKI AND W. J. BAUMOL, *The dual in nonlinear programming and its economic interpretation*, Rev. Economic Studies, XXXV, 103 (1968), pp. 237–256.

[2] C. BERGE AND A. GHOUILA-HOURI, *Programming, Games and Transportation Networks*, John Wiley, New York, 1965.

[3] S. P. BRADLEY, *Decomposition programming and economic planning*, Rep. 67-20, Operations Research Center, University of California, Berkeley, 1967.

[4] A. CHARNES, W. W. COOPER AND K. O. KORTANEK, *On the theory of semi-infinite programming and a generalization of the Kuhn-Tucker saddle point theorem for arbitrary convex functions*, Naval Res. Logist. Quart., 16 (1969), pp. 41–51.

[5] R. W. COTTLE, *Symmetric dual quadratic programs*, Quart. Appl. Math., 21 (1963), pp. 237–243.

[6] G. B. DANTZIG, *Linear Programming and Extensions*, Princeton University Press, Princeton, 1963.

[7] G. DEBREU, *Theory of Value*, John Wiley, New York, 1959.

[8] W. S. DORN, *Duality in quadratic programming*, Quart. Appl. Math., 18 (1960), pp. 155–162.

[9] H. EVERETT, *Generalized Lagrange multiplier method for solving problems of optimum allocation of resources*, Operations Res., 11 (1963), pp. 399–417.

[10] J. E. FALK, *A constrained Lagrangian approach to ·nonlinear programming*, Doctoral thesis, University of Michigan, Ann Arbor, 1965.

[11] W. FENCHEL, *Convex cones, sets, and functions*, Lecture Notes, Princeton University, Princeton, 1953.

[12] A. V. FIACCO AND G. P. MCCORMICK, *Nonlinear Programming*, John Wiley, New York, 1968.

[13] B. L. FOX, *Stability of the dual cutting-plane algorithm for concave programming*, RM-6147-PR, The RAND Corporation, Santa Monica, Calif., 1970.

[14] M. FRANK AND P. WOLFE, *An algorithm for quadratic programming*, Naval Res. Logist. Quart., 3 (1956), pp. 95–110.

[15] D. GALE, *A geometric duality theorem with economic applications*, Rev. Economic Studies, 34 (1967), pp. 19–24.

[16] A. M. GEOFFRION, *Primal resource-directive approaches for optimizing nonlinear decomposable systems*, Operations Res., 18 (1970), pp. 375–403.

[17] ———, *Elements of large-scale mathematical programming*, Management Sci., 16 (1970), pp. 652–691.

[18] ———, *Generalized Benders decomposition*, J. Optimization Theory Appl., to appear.

[19] E. G. GOLSHTEIN, *A general approach to the linear programming of block structures*, Soviet Physics Dokl., 11 (1966), no. 2, pp. 100–103.

[20] F. J. GOULD, *Nonlinear Duality Theorems*, Rep. 6817, Center for Mathematical Studies in Business and Economics, University of Chicago, Chicago, 1968.

[21] K. O. KORTANEK AND J. P. EVANS, *Asymptotic Lagrange regularity for pseudoconcave programming with weak constraint qualification*, Operations Res., 16 (1968), pp. 849–857.

[22] L. S. LASDON, *Duality and decomposition in mathematical programming*, IEEE Trans. System Sci. and Cybernetics, SSC-4 (1968), pp. 86–100.

[23] D. G. LUENBERGER, *Optimization by Vector Space Methods*, John Wiley, New York, 1969.

[24] O. L. MANGASARIAN, *Nonlinear Programming*, McGraw-Hill, New York, 1969.

[25] O. L. MANGASARIAN AND J. PONSTEIN, *Minmax and Duality in Nonlinear Programming*, J. Math. Anal. Appl., 11 (1965), pp. 504–518.

[26] R. T. ROCKAFELLAR, *Duality in nonlinear programming*, Mathematics of the Decision Sciences, Part 1, G. B. Dantzig and A. F. Veinott, eds., American Mathematical Society, Providence, 1968, pp. 401–422.

[27] ———, *Convex functions and duality in optimization problems and dynamics*, Mathematical Systems Theory and Economics, H. W. Kuhn and G. P. Szegö, eds., Springer-Verlag, Berlin, 1969.

[28] J. STOER, *Duality in nonlinear programming and the minimax theorem*, Numer. Math., 5 (1963), pp. 371–379.

[29] I. TAKAHASHI, *Variable separation principle for mathematical programming*, J. Operations Res. Soc. Japan, 6 (1964), pp. 82–105.

[30] H. Uzawa, *Iterative methods for concave programming*, Studies in Linear and Nonlinear Programming, K. Arrow, L. Hurwicz and H. Uzawa, eds., Stanford University Press, Stanford, Calif., 1958.

[31] R. M. Van Slyke and R. J. B. Wets, *A duality theory for abstract mathematical programs with applications to optimal control theory*, J. Math. Anal. Appl., 22 (1968), pp. 679–706.

[32] A. C. Williams, *Nonlinear Activity Analysis and Duality*, presented at the Sixth International Symposium on Mathematical Programming, Princeton, 1967.

[33] R. Wilson, *Computation of optimal controls*, J. Math. Anal. Appl., 14 (1966), pp. 77–82.

[34] P. Wolfe, *A duality theorem for non-linear programming*, Quart. Appl. Math., 19 (1961), pp. 239–244.

4. Mathematical Programming and Control Theory: Trends of Interplay[*]
David G. Luenberger[†]

I. Introduction

There is a common phenomenon in science for two initially distinct fields to progress independently for a number of years and then
later, while still maintaining their individual identity, find that
they share a common core of basic principles. Once the commonality
is recognized it is natural to expect a variety of attitudes and activities directed toward interaction between the fields. There will
be the skeptic who is certain that the other field is merely a special case of his own, and there will be the restless who will immediately turn to this other surely more exciting field. Ideally, there
will be cross-fertilization, which amounts to the application of techniques from one field to specific problems in the other. There will
be development of the common core of principles, derived from comparison and refinement of the principles already established by each
field separately, that ultimately forms the foundation of a more general discipline. Finally, there will be numerous survey articles
and lectures on the connections between the two fields.

The fields of mathematical programming and control theory have
developed in a fashion similar to that described above. We restrict
attention in this paper, however, to the areas of greatest overlap,
i.e. to certain subareas within each field. Thus, by mathematical
programming we refer here to continuous variable programming, including linear and nonlinear programming. We exclude dynamic programming, not because it is irrelevant but, contrarily, because its connection with control is so direct that it has equal claim by both
fields. By control theory we refer exclusively to optimal control,
leaving aside the general theory of dynamic systems (stability, con-
trollability, observability, etc.).

[*]This research was supported in part by the National Science Foundation under Grant NSF-GK-16125. Paper presented at the 7th International Symposium on Mathematical Programming, The Hague, The Nether-
lands, Sept. 1970.

[†]Stanford University.

General recognition, by researchers in both fields, of the commonality between the fields occurred roughly in about 1962, but naturally it was recognized by many individuals before then. Since that
time there has been vigorous interplay between the two fields and numerous conferences whose themes enhanced this interplay. Both fields
have, as a result, been substantially influenced by one another. On
balance, however, to date the contribution of mathematical programming to control theory is generally better recognized than the reverse contribution. This is due to the fact that mathematical programming, having less inherent structure, can from one viewpoint be
regarded as the more general--with control problems being a special
case--and to the fact that the modern development of mathematical
programming predated modern control theory by about seven years. On
the other side, however, the particular structure of control problems
introduces some difficulties and some simplifications that were not
conceived under the general programming setting. Resolution and exploitation of these has led to both theoretical and algorithmic development in mathematical programming.

Both mathematical programming and control can be naturally dichotomized into the study of properties of solution points, i.e. necessary and sufficient conditions, and the development and analysis of
effective computational methods for obtaining a solution. We adhere
to this natural dichotomy in the discussion presented in this paper.

Section II presents a brief account of the interplay between the
fields that was inspired by the initial gap between the primary sets
of necessary conditions for optimality: the Kuhn-Tucker Theorem for
mathematical programming, and the Pontryagin maximum principle for
control. Here we point out how the initial gap was closed by working
from both directions. Sections III and IV discuss computational
methods. In this area the contribution has been almost entirely
from mathematical programming to control, but there have been differences in emphasis and misinterpretations among the two groups
that has tended to retard full development in this area. We point
out some of these differences in viewpoint and offer some sugges-
tions for closing these gaps.

The paper is not intended to be a comprehensive survey of work in either field nor of the interaction between them (a discussion of al- gorithms alone would constitute a large volume). Instead, the paper attempts to point out how certain initial differences were satisfac- torily resolved, and attempts to bring into focus some issues that have not yet received full attention and which require further inter- active investigation. The selection of topics that are highlighted in this paper are, of course, somewhat influenced by the author's in- terest, areas of experience, and personal assessments as to where further interaction will be most fruitful.

II.   Optimality Conditions

2.1   Continuous-Time Problems

The Pontryagin Maximum Principle is the fundamental set of nec- essary conditions for continuous-time optimal control problems. We state a version of it here partly to display the result itself and partly for the purpose of introducing the continuous-time optimal control problem [1].

Continuous-Time Optimal Control Problem (Fixed Terminal Time)

Consider a dynamical system described by the system of ordinary differential equations

$$(2.1) \qquad \frac{d}{dt} x(t) = h\Big(x(t), u(t)\Big) \qquad t \in [0,T]$$

where $x(t) \in E^n$ is the state of the system at time t, $u(t) \in E^p$ is the control input at time t, and h takes values in E^n. Given an initial state $x_0 \in E^n$, a cost function f from $E^n \times E^p$ into the reals, a set $U \subset E^p$, and a terminal constraint function g from E^n into E^m, $m \leq n$, find the piecewise continuous control $u(\cdot)$ on $[0,T]$ and corresponding state trajectory $x(\cdot)$ deter- mined from (2.1) satisfying

$$(2.2a) \qquad u(t) \in U \qquad \text{for all} \quad t \in [0,T]$$

$$(2.2b) \qquad x(0) = x_0$$

(2.2c) $$g\Big(x(T)\Big) = 0$$

that minimizes the objective

(2.3) $$J = \int_0^T f\Big(x(t), u(t)\Big) dt .$$

There are alternative formulations of the continuous-time optimal control problem in which the objective is expressed as a function
of the terminal state and some auxiliary constraints are expressed
as constraints on the value of an integral involving the state and
control. These alternative formulations are, under mild smoothness
assumptions, all equivalent. One formulation can be transformed to
another by introduction of appropriate additional state variables and
associated augmentation of the system equations. These transformations are now quite familiar to control theorists and we shall not
belabor them here, but make free implicit use of them throughout the
paper.

To formulate the maximum principle we introduce the additional
assumptions that the functions f and h are continuously differentiable with respect to x and continuous with respect to u and
that the function g is continuously differentiable, and its Jacobian m × n matrix has rank m for all x satisfying $g(x) = 0$.

Pontryagin Maximum Principle

If $\hat{u}(\cdot)$ together with its associated state trajectory $\hat{x}(\cdot)$
is a solution to the fixed terminal time optimal control problem,
then there exist an adjoint trajectory $\lambda(\cdot)$ from $[0,T]$ into E^n,
a $\lambda^0 \leq 0$, and a $\mu \in E^m$ such that $(\lambda^0, \mu) \in E^{m+1}$ is not zero and

(2.4) $$-\frac{d}{dt}\lambda(t) = \left[\frac{\partial h}{\partial x}\Big(\hat{x}(t), \hat{u}(t)\Big)\right]' \lambda(t) + \lambda^0 \left[\frac{\partial f}{\partial x}\Big(\hat{x}(t), \hat{u}(t)\Big)\right]'$$

for $t \in [0,T]$,

(2.5) $$\lambda(T) = \left[\frac{\partial g}{\partial x}\Big(\hat{x}(T)\Big)\right]' \mu$$

and such that for almost all $t \in [0,T]$, the Hamiltonian

$$(2.6) \qquad H\left(\hat{x}(t), \lambda(t), v\right) = \lambda^{o} f\left(\hat{x}(t), v\right) + \lambda'(t) h\left(\hat{x}(t), v\right)$$

is maximized by $\hat{u}(t)$, i.e. for every $v \in U$

$$(2.7) \quad \lambda^{o} f\left(\hat{x}(t), \hat{u}(t)\right) + \lambda'(t) h\left(\hat{x}(t), \hat{u}(t)\right) \geq \lambda^{o} f\left(\hat{x}(t), v\right)$$
$$+ \lambda'(t) h\left(\hat{x}(t), v\right).$$

An unusual feature of this principle, over that of more classical necessary conditions, is that they are partly expressed in differential form, as in (2.4), and partly in global form, as in (2.7).
This mixture of differential and global Lagrange multiplier conditions
is, as discussed later in this section, a challenging point even in
a finite-dimensional setting.

The multiplier $\lambda^{o} \leq 0$ is of special concern for if it is equal
to zero then the objective does not enter the conditions. A non-de-
generate problem will have $\lambda^{o} < 0$, in which case it is set to
$\lambda^{o} = -1$, but we shall not pursue the conditions that guarantee this.

The original proof of the maximum principal given in [1] is
quite involved, closely following classical arguments in the calculus
of variations and physics, where the term Hamiltonian was borrowed,
and bearing little relation to work in mathematical programming. Indeed, for a few years it was the relations between the calculus of
variations and continuous-time control that highlighted work on necessary conditions--mathematical programming being excluded as being
restricted to finite-dimensional problems.

Attention quickly was directed, both in mathematical programming
and control circles, to an attempt to generate a parallel result for
discrete time optimal control problems--the former circle because the
resulting problem is finite-dimensional, the latter because sampled
systems were becoming increasingly important as computers found their
way into control applications. This was the first strong point of
interaction between the fields.

2.2 Discrete-Time Problems

<u>Discrete Time Optimal Control Problem (Fixed Terminal Time)</u>

Consider a dynamical system described by the system of difference equations

$$(2.8) \qquad x(k+1) - x(k) = h\Big(x(k), u(k)\Big) \quad , \quad 0 \le k \le N-1$$

where $x(k) \in E^n$, $u(k) \in E^p$, and h maps into E^n. Given an initial state $x_0 \in E^n$, f mapping from $E^n \times E^p$ into the reals, a set $U \subset E^p$, and g mapping from E^n into E^m, find the sequence of controls $u(k)$, $k = 0, 1, \ldots, N-1$ and corresponding state trajectory determined from (2.8) satisfying

$$(2.9a) \qquad\qquad u(k) \in U \quad , \quad 0 \le k \le N-1$$

$$(2.9b) \qquad\qquad x(0) = x_0$$

$$(2.9c) \qquad\qquad g\Big(x(N)\Big) = 0$$

that minimizes the objective

$$(2.10) \qquad\qquad J = \sum_{k=0}^{N-1} f\Big(x(k), u(k)\Big) .$$

Again it is assumed that f and h are continuously differentiable with respect to x but only continuous with respect to u; and that the Jacobian matrix of g has rank m for all x satisfying $g(x) = 0$ [2].

The problem of obtaining a satisfactory result for the discrete-time problem was soon found to be extremely challenging and was taken up by both mathematical programming and control theorists. Available results for finite-dimensional problems, such as the Kuhn-Tucker Theorem, were able to yield results of the mixed differential and global form only for the most simple problem structures (e.g. linear dynamics, convex costs, and convex constraints); hence this problem motivated work directed toward extending these classical results. A notable contribution in this respect is a generalized Fritz John result [3], [4] which yields a maximum principle result for the discrete-time problem.

There is no unique way in which to characterize inherent

difference between discrete-time problems and continuous-time problems which leads to the analytic difficulty of obtaining a maximum
principle. From the viewpoint of more or less classical variational
theory the difference can be traced to the fact that in continuous-
time a control history perturbation of large magnitude but arbitrarily short duration can be considered which introduces only small
changes in the resulting trajectory; while in discrete-time a large
magnitude perturbation of control at any instant introduces a large
change in the trajectory. Thus, only in continuous-time is it possible to introduce control perturbations that are simultaneously
large in magnitude (at a given instant) but small in their effect on
the objective and constraint functionals. Alternatively, from a
viewpoint more closely connected with mathematical programming, the
difference between continuous-time and discrete-time problems is that
a certain set of trajectory end-points, defined by allowing a family
of perturbations, is convex in the continuous-time case but not necessarily convex in the discrete-time case.

From a control theoretic viewpoint (i.e. by explicit consideration of trajectories) Halkin [5] obtained a mixed differential-global
maximum principle for the discrete-time problem that exploits a convexity assumption. His result was slightly strengthened by Holtzman [6] who pointed out that convexity could be replaced by directional convexity, a concept closely related, as we point out later
in this section, to that of a convex epigraph popular in mathematical programming.

While work progressed on the discrete-time problem two other
related activities were also, at least partially, inspired by the
maximum principle. The first was the further development of a general theory of optimization valid in infinite-dimensional spaces.
Here the aim was to obtain results having the simple structural char-
acteristics of the classical programming results but wider applicability. This theory is characterized by the work of Liusternik [7],
Hurwicz [8], Rockafellar [9], Luenberger [10], and many others [11-
15]. The second activity, initiated by Gamkrelidze [16] and developed by Neustadt [17-19], Halkin [20], and Canon, Cullum, and Polak
[2], [21], was aimed more directly toward the maximum principle.

This took the form of a generalized theory of optimization for problems having a finite number of equality constraints from which the Pontryagin maximum principle (discrete or continuous-time) as well as the classical mathematical programming results are easily deduced.

It is impossible to obtain a maximum principle for the discrete-time optimal control problem that is as strong as that for the corresponding continuous-time problem. One must settle for a weaker condition than global maximization of the Hamiltonian over U, or some sort of convexity assumption must be introduced to yield the maximization condition. We state here one version that closely parallels the original one for continuous-time problems[2].

Discrete-Time Maximum Principle

Assume that for every $x \in X^n$ the set in E^{n+1} defined by

$$\{(r,z): \quad r = f(x,u), \quad z = h(x,u), \quad u \in U\}$$

is $(-1,0,0,\ldots,0)$-directionally convex. (See below).

If $\hat{u}(\cdot)$ together with its associated state trajectory $\hat{x}(\cdot)$ is a solution to the discrete-time optimal control problem, then there exist an adjoint trajectory $\lambda(k) \in E^n$, $k = 0,1,2,\ldots,N$, a $\lambda^o \leq 0$, and a $\mu \in E^m$, such that $(\lambda^o,\mu) \in E^{m+1}$ is not zero and

$$(2.11) \quad \lambda(k) - \lambda(k+1) = \left[\frac{\partial h}{\partial x}\left(\hat{x}(k),\hat{u}(k)\right)\right]'\lambda(k+1) + \lambda^o\left[\frac{\partial f}{\partial x}\left(\hat{x}(k),\hat{u}(k)\right)\right]'$$

for $k = 0,1,2,\ldots,N-1$,

$$(2.12) \qquad\qquad \lambda(N) = \left[\frac{\partial g}{\partial x}\left(\hat{x}(N)\right)\right]'\mu$$

and such that for every $v \in U$ and $k = 0,\ldots,N-1$,

$$(2.13) \quad \lambda^o f\left(\hat{x}(k),\hat{u}(k)\right) + \lambda'(k)h\left(\hat{x}(k),\hat{u}(k)\right) \geq \lambda^o f\left(\hat{x}(k),v\right)$$

$$+ \lambda'(k)h\left(\hat{x}(k),v\right) .$$

2.3 Local-Global Results

In the remainder of this discussion of optimality conditions we turn to the definition of directional convexity, its relation to epigraphs, and to an elaboration upon the theme most striking about the original maximum principle--the categorization of the problem variables into local and global parts. We show that this theme can easily be incorporated into the general theory of optimization.

A set S in a vector space V is said to be <u>e-directionally convex</u>, where $e \in V$, if for every vector v in the convex hull of S there is a vector $s \in S$ and $\beta \geq 0$ such that $v = s + \beta e$.

An equivalent and perhaps simpler definition is that S is e-directionally convex if the set $\Omega = \{v: \ v = s - \beta e, \ \beta \geq 0, \ s \in S\}$ is convex. In other words, if the sum of S and the ray $-\beta e$ is a convex set. If f is a functional on a convex subset Ω , then f is convex if its graph in $R \times V$ defined by $\{(r,v): \ r = f(v),$ $v \in \Omega\}$ is $(-1,0)$-directionally convex, which is equivalent to f having a convex epigraph.

We conclude this section with a theorem that is not difficult to prove but which captures the spirit of the original maximum principle and transfers it to mathematical programming. The proof itself employs ideas used frequently in mathematical programming and thus also captures the spirit of that field. In order not to introduce new concepts the result is stated here in finite-dimensional form, but is easily extended to situations where both the unknowns and the constraints are infinite dimensional. The result is, in a sense, the analog for inequalities of the general results obtained by Neustadt and others [16-21] for problems with a finite number of equalities. The generalized inequality result, however, is easier to prove, is not restricted to a finite number of constraints, and carries with it a regularity assumption that guarantees a unity coefficient on the objective in the Lagrangian.

Local-Global Programming Problem

Find x and u to

(2.13a)   minimize $f(x,u)$

(2.13b)   subject to $g(x,u) \leq 0$

(2.13c) $u \in \Omega$

where $x \in E^n$, $\Omega \subset E^p$ is convex, and the functions f and g take
values in the reals and E^p respectively, and both are continuously
differentiable with respect to x. For each fixed x the set in
E^{p+1}

$$\{(r,z): \quad r = f(x,u), \quad z \geq g(x,u), \quad u \in \Omega\}$$

is $(-1,0,0,\ldots,0)$-directionally convex.

This problem possesses standard differentiability properties
with respect to x and convexity properties with respect to u. Indeed, if f and g are both convex with respect to u for each
fixed x, the directional convexity assumption will be satisfied,
and for fixed x the problem is a convex program in u. We introduce a regularity condition that is a natural combination of standard
conditions for local and global theories [10]--that if x_0, u_0 is a
solution, then there is a $h_0 \in E^n$, $v_0 \in \Omega$ such that

(2.14) $g(x_0,v_0) + \nabla_x g(x_0,u_0)h_0 < 0$.

Lagrange-Multiplier Result

Let x_0, u_0 be an optimal solution to the local-global program (2.13) and suppose the solution satisfies the regularity condition (2.14). Then there is a $\lambda \in E^p$, $\lambda \geq 0$ such that

(2.15) $\nabla_x f(x_0,u_0) + \lambda' \nabla_x g(x_0,u_0) = 0$,

(2.16) $f(x_0,u_0) + \lambda' g(x_0,u_0) \leq f(x_0,v) + \lambda' g(x_0,v)$,

for all $v \in \Omega$, and

(2.17) $\lambda' g(x_0,u_0) = 0$.

<u>Proof.</u>

   Define the mapping G taking values in E^{p+1} by

$$G(x,u) = \Big(f(x,u) - f(x_0,u_0), \quad g(x,u)\Big) .$$

In E^{p+1} defines the sets

$$A = \{y: \ y \geq G(x_0,v) + \nabla_x G(x_0,u_0)h, \quad v \in \Omega\}$$

$$B = \{y: \ y \leq 0\} .$$

Both of these sets are convex--A because of the directional convexity assumption--and B has nonempty interior.

   We now show that A contains no interior points of B. Suppose there is a point in A which is interior to B. This means
there is $h \in E^n$, $v \in \Omega$ such that

$$G(x_0,v) + \nabla_x G(x_0,u_0)h < 0 .$$

The point $G(x_0,v) + \nabla_x G(x_0,u_0)h$ is the center of some open sphere
contained in the negative orthant N of E^{p+1}. Suppose this sphere
has radius $\rho > 0$. Then $\alpha[G(x_0,v) + \nabla_x G(x_0,u_0)h]$ is the center of
such a sphere having radius $\alpha\rho$; hence so is the point

$$(1-\alpha)G(x_0,u_0) + \alpha[G(x_0,v) + \nabla_x G(x_0,u_0)h] .$$

This point can be written

$$G(x_0,u_0) + \alpha\nabla_x G(x_0,u_0)h + \alpha[G(x_0,v) - G(x_0,u_0)]$$

$$= G(x_0 + \alpha h,u_0) + \alpha[G(x_0,v) - G(x_0,u_0)] + o(\alpha)$$

$$= (1-\alpha)G(x_0 + \alpha h,u_0) + \alpha G(x_0 + \alpha h,v)$$

$$+ \alpha[G(x_0 + \alpha h,u_0) - G(x_0,u_0)]$$

$$- \alpha[G(x_0 + \alpha h,v) - G(x_0,v)] + o(\alpha)$$

$$= (1-\alpha)G(x_0 + \alpha h,u_0) + \alpha G(x_0 + \alpha h,v) + o(\alpha) .$$

By the assumption of directional convexity there is a point $v_\alpha \in \Omega$ such that

$$G(x_0 + \alpha h, v_\alpha) \leq (1-\alpha)G(x_0 + \alpha h, u_0) + \alpha G(x_0 + \alpha h, v) .$$

Thus the point $G(x_0 + \alpha h, v_\alpha) + o(\alpha)$ is contained in a sphere of radius $\alpha \rho$ in the negative orthant. Hence, for sufficiently small α, $G(x_0 + \alpha h, v_\alpha) < 0$. This contradicts the optimality of x_0, u_0 and hence A contains no interior points of B.

In view of the above conclusions there is a hyperplane separating the sets A and B. Thus, there is a vector in E^{p+1} of the form $\psi = (r, \lambda)$ with r real, $\lambda \in E^p$, and a δ such that

$$\psi' y \geq \delta \quad \text{for all} \quad y \in A$$

$$\psi' y \leq \delta \quad \text{for all} \quad y \in B .$$

We take $\delta = 0$ since the point $(0,0)$ belongs to both A and B. Since B is the negative orthant it follows immediately that $\psi \geq 0$ or equivalently that $r \geq 0$, $\lambda \geq 0$. Furthermore, $r \neq 0$ because otherwise the hyperplane would not separate the point $G(x_0, v_0) + \nabla_x G(x_0, u_0) h_0$ from B. Hence, without loss of generality we may take $r = 1$.

From the separation property we must have

$$\psi' G(x_0, u_0) \geq 0$$

and thus

$$\lambda' g(x_0, u_0) \geq 0$$

but since

$$\lambda \geq 0, \quad g(x_0, u_0) \leq 0$$

it follows that

$$\lambda' g(x_0, u_0) = 0 .$$

From the separation property for A we have

$$f(x_0,v) - f(x_0,u_0) + \lambda'g(x_0,v) + \nabla_x f(x_0,u_0) h + \lambda'\nabla_x g(x_0,u_0)h \geq 0 .$$

Setting $h = 0$ and using (2.17) yields (2.15). Noting that h is arbitrary yields (2.16).

III. Computation (Continuous-Time Problems)

3.1 Evaluation of Gradient

Consider, initially, the following simple structure.

Continuous-Time Optimal Control

Given a dynamical system

(3.1)
$$\frac{d}{dt} x(t) = h\Big(x(t),u(t)\Big)$$

together with initial condition $x(0) = x_0$ where $x(t) \in E^n$, $u(t) \in E^p$, $t \in [0,T]$ and given the objective functional

(3.2)
$$J = \int_0^T f\Big(x(t),u(t)\Big) dt$$

where f and h are both continuously differentiable with respect to x and u, find $u(\cdot)$ to minimize J.

This structure is the simplest nontrivial one for control since it is unconstrained except for the dynamical relation between $x(\cdot)$ and $u(\cdot)$. Indeed, the most profitable manner in which to view this problem is as an unconstrained problem with respect to $u(\cdot)$. Note that if $u(\cdot)$ is specified on $[0,T]$ then $x(\cdot)$ is determined uniquely by the dynamic system (3.1) and the initial conditions. This means that the value of the objective functional is also uniquely determined. Thus, we often write $J(u)$, rather than simply J, to explicitly point out this implicit dependence.

To obtain necessary conditions for a problem with this structure or to formulate an algorithmic scheme for its solution, it is most natural to seek an expression for the gradient of J (with respect to u). Thus given $u(\cdot)$ and its associated $x(\cdot)$ we have

(3.3) $$\nabla_u J(u) = \lambda(t)'\nabla_u h\Big(x(t),u(t)\Big) + \nabla_u f\Big(x(t),u(t)\Big)$$

where

$$(3.4a) \qquad - \frac{d}{dt} \lambda(t) = \left[\nabla_x h\big(x(t), u(t)\big) \right]' \lambda(t) + \nabla_x f\big(x(t), u(t)\big)'$$

$$(3.4b) \qquad\qquad\qquad \lambda(T) = 0 .$$

Necessary conditions are obtained by setting this gradient to zero.[*] For computation we note that given $u(\cdot)$, forward integration of the system equations (3.1) yields $x(\cdot)$. Then backward integration of the adjoint equations (3.4) yields $\lambda(\cdot)$, and the gradient can be directly evaluated from (3.3). Additional constraints are incorporated in the necessary conditions through introduction of Lagrange multipliers and in computational techniques by gradient projection, etc.

This procedure for evaluation of the gradient forms the backbone of most algorithms in use designed to solve control problems. It can be regarded as the basic technique for exploiting the dynamic structure of the problem.

3.2 Direct Approach for Continuous-Time Problem

The usual approach taken by control theorists when solving a continuous-time problem is to carry out analyses and develop algorithms in continuous time and then execute the required computations as accurately as necessary on a digital computer. This viewpoint more or less excludes consideration of simplex-like algorithms since they are not applicable to infinite-dimensional problems but does not exclude the numerous programming techniques that can be generalized to infinite dimensions.

There have been a number of distinct techniques suggested for the continuous-time problem, essentially all of which parallel techniques developed for general nonlinear programming problems. One class of techniques are those that attempt to solve the equations

---

[*] The adjoint variable $\lambda(\cdot)$ resulting from (3.4) is the negative of that in the Pontryagin formulation. The form (3.4) seems, from many respects, the most natural, and it has been argued that things would have been simpler if Pontryagin had used (3.4) and formulated a "minimum principle." As it is, both formulations of the adjoint equation occur with about equal frequency in the literature.

expressing the necessary conditions [22]. In this approach $u(t)$ is expressed in terms of $\lambda(t)$ and $x(t)$ from (2.7). The remaining unknowns $x(\cdot)$, $\lambda(\cdot)$ then satisfy a system of $2n$ ordinary differential equations with boundary conditions at both the initial and final times. This reduces the problem to that of numerically solving a two-point boundary value problem. Both first-order (simple successive approximation) and second-order (Newton-type) methods have been developed to solve the resulting two-point boundary value problem.

Another technique, called the epsilon method [23], is to consider the dynamic system equation as a differential constraint and treat it by a penalty function technique. Generally, this yields an ill-conditioned problem. If, however, the problem has special analytical structure and if a second order minimization routine can be used, the technique is workable.

By far the most widely used methods are descent methods such as the gradient method [24], [25], conjugate gradients [26], [27], and Newton's method [28-31]. All of these have solved a variety of optimal control problems with good success.

The success of gradient methods lies, as pointed out above, in the simple procedure available for evaluation of the gradient in a control problem. There is a parallel to this for Newton's method that makes second-order methods computationally feasible. The second-order correction is found by integrating a matrix Riccati equation, of dimension $n \times n$, backward from T to the initial time. Thus, the effort to obtain the second-order correction is only about n times that of obtaining the gradient. Details can be found in [30]. The discrete-time analog of the Riccati equation is discussed in Section 4.2 below.

The precise details associated with these methods, step size selection, incorporation of constraints, etc., are largely identical to those developed for general nonlinear programming. Thus, although this is an area of major cross fertilization between mathematical programming and control (mainly from the former to the latter), it is senseless to attempt to detail these procedures here.

3.3   Free Terminal Time Problems

In the free terminal time problem the final time   T   is not
specified a priori but is determined as part of the optimization.
In problems of this type there is usually a terminal constraint on
the state vector that implicitly determines the final time for a
given trajectory.   This is exemplified by the problem of guiding a
rocket to the moon with minimum fuel expenditure; or the aircraft
climb problem where it is desired to control the angle of attack of
an aircraft so as to reach a given altitude in minimum time.   The
terminal time associated with a trajectory is determined by the terminal constraint.

Problems of this type cannot directly be cast into the framework
of a vector space defined over   [0,T]   as is done, at least implicitly, in fixed-time problems.   Several techniques have been employed
to handle the problem indirectly.   One, often used in conjunction
with good physical insight, is to reparameterize the equations of
motion with respect to the terminal constraint variable.   In the aircraft climb problem, for example, the dynamic equations might be re-
parameterized with altitude as the independent running variable, instead of time, so that the problem is transformed into one on a fixed
interval.   This scheme is most effective only if the terminal constraint can be expressed in terms of a variable that itself can be
expected to increase along a trajectory.

Another approach to variable time problems, a favorite among
mathematical programmers, is to solve, for various values of terminal
time   T,   the associated problem having fixed terminal time   T.   Successive trial values of   T   can be generated by a curve fitting procedure applied to the points of cost versus final time.   Logically,
this technique amounts to having an inner loop that minimizes with
respect to   u(·)   and an outer loop that minimizes with respect to T.

We suggest here that problems with variable terminal time can
often be most effectively solved through the introduction of a penalty function, which in these situations not only eliminates the
terminal constraint by modifying the objective but also allows one to
consider the problem as though it were of fixed duration.

We write the problem as

$$\min_{u(\cdot),T} \int_0^T f\Big(x(t),u(t)\Big)\,dt$$

(3.5)   subject to $\dot{x}(t) = h\Big(x(t),u(t)\Big)$, $x(0) = x_0$

$$g\Big(x(T)\Big) = 0$$

and cast the problem into vector space format by considering that
$u(\cdot)$ is defined on an interval $[0,T_1]$ where T_1 is sufficiently
large to contain any interval of interest. We then solve the approximating penalty problem

(3.6) $$\min_{u(\cdot),x(\cdot)} \min_T \int_0^T f\Big(x(t),u(t)\Big)\,dt + K\|g\Big(x(T)\Big)\|^2$$

where

$$\dot{x}(t) = h\Big(x(t),u(t)\Big) \quad , \quad x(0) = x_0$$

and K is some (large) positive constant.

The minimization problem is unconstrained and can be regarded
as being one of finding $u(\cdot)$ on $[0,T_1]$ with T being determined
from $u(\cdot)$ by minimization along the resulting trajectory. The problem is thus a standard fixed-time constraint-free problem, except
that the objective function does not have the usual integral form but
must be evaluated by minimization over T--a computationally minor
modification since the integral is automatically generated as a function of T. Furthermore, the gradient of this objective can be calculated by backward integration from the current T just as in fixed
time problems [32].

3.4 Discretization of Continuous-Time Problems

When employing a digital computer to solve a continuous-time optimal control problem, operations such as the solution of differential equations cannot be executed exactly. This means that either
the original problem itself or its solution (or both) must be approximated in some appropriate fashion. Aside from some fairly specialized function approximation schemes, there are two primary approaches

to this approximation problem--both of which are based on discretization of time. Crudely, the approaches can be labeled the mathematical programming and the control theory approaches--but these labels are not meant to have any real significance; they are based on the observation that workers from each field have tended to favor one particular approach to the discretization problem.

The first approach, that favored by mathematical programmers, is to discretize the original problem at the outset, converting it to a discrete-time optimal control problem, and then later direct attention toward the solution of the resulting finite-dimensional problem. The simplest scheme for doing this is to define $\Delta = \frac{T}{N}$ for some positive integer N and introduce the notation $x(k) \equiv x(k\Delta)$, $u(k) \equiv u(k\Delta)$. Next, define the difference equations

$$(3.7) \qquad \frac{x(k+1) - x(k)}{\Delta} = h\Big(x(k),u(k)\Big) \quad , \quad 0 \leq k < N$$

and the objective functional

$$(3.8) \qquad J = \sum_{k=0}^{N-1} f\Big(x(k),u(k)\Big) \, ,$$

which approximate the system differential equations and objective respectively. If the original problem has additional constraints, they are discretized in a similar manner.

There are a number of important questions regarding the appropriateness of such approximations in terms of their convergence to the true solution as Δ goes to zero. For information on this topic see Cullum [33] and Daniel [34]. Roughly, however, the accuracy can be expected to be on the order of Δ, i.e. $O(\Delta)$. In some circumstances, of course, it may be advantageous to use nonuniform discretization intervals.

This first approach to discretization is attractive philosophically because the approximation is made at the outset and then an accurate answer to the approximate problem can be sought. It is attractive to a mathematical programmer because the whole array of programming techniques becomes applicable to the approximate problem.

Unfortunately, it can easily get out of hand yielding a problem of enormously large dimension with poor accuracy properties.

The second approach, the one largely employed by control theorists, is based on analysis of the continuous-time problem. The problem is constantly regarded as being one in continuous time and one attempts to carry out, at least approximately, the calculations dictated by the continuous time analysis. The primary components of these calculations is integration of the system differential equations forward from $t = 0$ to $t = T$ given a control history over this interval and integration of the adjoint equations backward to obtain the gradient. Again, this is done by consideration of N equally spaced points in the interval $[0,T]$, but the first-order Euler approximation is generally found to be too crude for practical use--or more precisely, Δ must be taken extremely small to yield a satisfactory approximation. It is more efficient to use higher order integration schemes such as a predictor-corrector, or Runge-Kutta procedure [35], [36].

For illustration consider the system

$$(3.9) \qquad \frac{d}{dt} x(t) = h\left(x(t),u(t)\right) .$$

We lay out intervals of equal width Δ and use the correspondence $x(k) \equiv x(k\Delta)$ as before. A standard corrector formula, such as the Adams-Moulton method, has the form

$$(3.10) \quad x(k+1)-x(k) = \Delta \sum_{i=k-q_k}^{k+1} \beta_{ik} h\left(x(i),u(i)\right) , \quad k = 0,1,2,\ldots,N\text{-}1 .$$

Usually, q_k is a constant, say q, for all values of k except near $k = 0$. To initialize the process q values $x(0), x(1),\ldots, x(q\text{-}1)$ must be generated by some other method, or we can develop a scheme with $q_k = k$ for $k \le q$. The value $q = 4$ is often used for the main part of the procedure.

The right-hand side of the corrector formula (3.10) depends on $x(k+1)$ and hence (3.10) cannot be used to determine $x(k+1)$ without iteration. For this reason a predictor formula, similar to

(3.10) but with different coefficients and without x(k+1) on the right-hand side, is used to obtain an initial estimate of x(k+1). Then starting from this estimate the corrector is solved by successive approximation.

Typically, when solving optimal control problems, the same integration procedure is used both for forward integration of the system equations and for backward integration of the adjoint system used for calculation of the gradient. The procedure is also used, in the case where a second-order procedure is being incorporated, for the backward integration of the matrix Riccati equation. This control theoretic approach to discretization has found wide acceptance because of its convenience and high accuracy, and because it fully exploits the simple method for generating the gradient.

It is impossible to muster a definitive argument that favors either of these two approaches over the other. Which is better is partially dependent on the overall problem structure beyond that of the dynamics alone. For the majority of continuous-time problems, however, the control theoretic viewpoint has a great advantage because of the relatively short time required for accurate integration. It is only problems having a plethora of constraints on the state vector and the control, or which have a particularly simple structure, such as being linear or quadratic programs (and for which consequently, gradient-based schemes must be replaced or supplemented by simplex-type algorithms) that the mathematical programming approach is superior.

We suggest here that it is possible to incorporate the best features of both of the above viewpoints into a single philosophically satisfying and computationally effective discretization viewpoint. Like the mathematical programmer, we decide that it is best to discretize the original problem into an approximating finite-dimensional one that can be attacked directly. Like the control theorist, however, we recognize that simple Euler discretization is too crude and that a more sophisticated approach must be taken. In order to make our discussion somewhat specific let us agree that a standard corrector formula with fixed step size Δ is an appropriate approximation to both the dynamic system and to the evaluation of the cost

function.   This, in turn, will produce a discretization of the problem of the form

(3.11) $X(k+1) = \Phi\Big(X(k),U(k)\Big)$

(3.12) $J = c^T x(N)$

where the dimensions of $X(k)$ and $U(k)$ are greater than n and p
by factors that depend on the particular corrector formula employed.
For convenience in what follows, we have assumed that the problem
has been put in the form of a terminal cost problem.

Philosophically, introduction of the corrector formula can be
viewed as a transformation of the original problem to a discrete-
time control problem.   Practically, however, this does not allow us
to bring into play all the algorithms applicable to most discrete problems, since the Φ function is made up of implicit functions that
can only effectively be evaluated by forward recursion in conjunction with an associated predictor.

Holding to the viewpoint that the discrete problem obtained
through the somewhat indirect process of using a predictor-corrector
scheme should be solved exactly, we next consider the question of
evaluating the associated gradient.   In the pure control theoretic
approach the gradient is approximated by applying the predictor-
corrector in reverse time to evaluate λ.   We instead consider the
possibility of evaluating this gradient exactly for the discrete
formulation obtained.   Evaluation of the gradient then forms the
basis for a family of optimization techniques.

Suppose that the optimal control problem is expressed as

(3.13)   minimize $J = c^T x(N)$

where

$$(3.14) \quad x(k+1)-x(k) = \Delta \sum_{i=k-q_k}^{k+1} \beta_{ik} h\Big(x(i),u(i)\Big) , \quad k = 0,1,2,\ldots,N-1 ,$$

$$x(0) = x_0.$$

To find the gradient of J with respect to $u(\cdot)$, multiply (3.14) by $\lambda(k)$, sum over k, and evaluate the total differential of the result, obtaining

$$\sum_{k=0}^{N-1} \lambda(k)dx(k+1) - \lambda(k)dx(k) = \sum_{k=0}^{N-1} \Delta \sum_{i=k-q_k}^{k+1} \beta_{ik}\lambda(k)\left[h_x\Big(x(i),u(i)\Big)dx(i)\right.$$

$$\left. + h_u\Big(x(i),u(i)\Big)du(i)\right] .$$

Or upon rearrangement,

$$\lambda(N-1)dx(N) + \sum_{i=1}^{N-1} [\lambda(i-1)-\lambda(i)]dx(i)$$

$$= \Delta \sum_{i=1}^{N-1} \left\{\sum_{k} \beta_{ik}\lambda(k)\right\}\left[h_x\Big(x(i),u(i)\Big)dx(i)\right.$$

$$\left. + h_u\Big(x(i),u(i)\Big)du(i)\right] .$$

Thus defining the λ sequence by

$$(3.15) \quad \lambda(i-1)-\lambda(i) = \Delta \sum_{k} \beta_{ik}\lambda(k)h_x\Big(x(i),u(i)\Big) , \quad i = 1,2,\ldots,N-1$$

$$\lambda(N-1) = c$$

$$dJ = \lambda(N-1)dx(N) = \Delta \sum_{i}\left[\sum_{k} \beta_{ik}\lambda(k)\right]h_u\Big(x(i),u(i)\Big)du(i)$$

and hence

$$(3.16) \qquad (\nabla J)_i = \Delta \left[\sum_{k} \beta_{ik}\lambda(k)\right]h_u\Big(x(i),u(i)\Big) .$$

This formula is similar to the adjoint equation used before, but it is not calculated by direct application of the corrector formula to the adjoint differential equation, as is done in the pure control theoretic approach. In the form presented here the λ sequence is evaluated by the simple backward formula (3.15) and then $\sum_{k} \beta_{ik}\lambda(k)$ is evaluated. In practice, it may be desirable to develop the appropriate recursion for $w(i) = \sum_{k} \beta_{ik}\lambda(k)$ and solve for it directly in a single backward sweep.

IV. Computation (Discrete-Time Problems)

4.1 General Remarks

If a control problem is inherently one in discrete time or if there are compelling reasons for using the Euler discretization of a continuous-time problem, then it takes the form

$$x(k+1) = \phi\Big(x(k),u(k)\Big) \ , \quad k = 0,1,2,\ldots,N-1$$

$$x(0) = x_0$$

$$J = \sum_{k=0}^{N-1} f\Big(x(k),u(k)\Big) + \Gamma\Big(x(N)\Big)$$

together with possibly several constraints restricting $x(k)$ and $u(k)$, $k = 0,\ldots,N$. Since the problem arose directly in this form, presumably the functions f and ϕ and the constraint functions are known explicitly. In this case the problem can be regarded as a finite-dimensional mathematical programming problem and the full range of algorithms becomes available.

There has been a good deal of computational experience accumulated on problems of this type which, on the whole, has essentially verified that general nonlinear programming algorithms can effectively solve control problems. Indeed, control problems often serve as convenient sources of large dimensional test problems for new algorithms. Those algorithms which use gradients can, of course, exploit the simple formula for the gradient of a discrete-time control problem. Generally, however, these straightforward applications of known algorithms to control problems carry little significance toward the fundamental development of either mathematical programming or optimal control. It is only when the special structure of large control problems is exploited in the development of an algorithm that there is a meaningful contribution.

A special case that is obviously potentially rich in structure is the linear problem, where the system equations, objective functional, and constraints are all linear. Some linear programs arising in this way have a natural structure that can be exploited to yield algorithms that are more efficient than the standard simplex method. A notable contribution in this respect is the work initiated by Dantzig [37] on dynamic Leontief models.

Another potentially rich special structure is where the system equations and constraints are linear and the objective function is quadratic (in both x and u). This leads to a quadratic programming problem that can be solved by standard algorithms. The dominant structure of these problems is often due to the dynamic equations rather than additional constraints and an efficient algorithm would fully exploit this. Indeed, the solution to the "linear system, quadratic cost" problem is one of the fundamental results from control theory.

4.2 Quadratic Problems

In the absence of additional constraints a simple form of the quadratic cost control problem is

(4.1a) $x(k+1) = \Phi x(k) + Bu(k)$, $k = 0,1,\ldots,N-1$

(4.1b) $x(0) = x_0$

(4.2) $J = \frac{1}{2} \sum_{k=1}^{N} x(k)'Qx(k) + u(k-1)'Ru(k-1)$

where Q and R are respectively $n \times n$ and $p \times p$ positive semidefinite matrices with either Q or R positive definite. Recalling that J may be regarded as a function of the sequence $u(\cdot)$, with the $x(\cdot)$ sequence determined from (4.1), it is clear that in this case J is a quadratic function of $u(\cdot)$. From a mathematical programming viewpoint the problem is equivalent to one of the form

$$\min\ USU - 2AU$$

where U is the control sequence vector of dimension p·N and S
is the induced p·N × p·N matrix. The solution is $U = S^{-1}A$ which
is trivial but unusable because S is not given explicitly and has
very high dimension. The special solution available for this problem
can thus be regarded as a method for exploiting the special structure
of S. Indeed, if one pursues the details of constructing S, the
special recursive solution available can be viewed as a special formula for solving a system of equations whose coefficient matrix is
block tridiagonal. We do not state the recursive solution for this
problem here since a more general problem is treated below.

Constrained Quadratic Cost Control Problem

Given the discrete dynamic equations

(4.3a) $x(k+1) = \Phi x(k) + Bu(k)$, $k = 0,1,2,\ldots,N-1$,

the initial condition

(4.3b) $x(0) = x_0$,

the objective

(4.4) $J = \frac{1}{2} \sum_{k=1}^{N} x(k)'Qx(k) + u(k-1)'Ru(k-1)$,

and the constraints

(4.5) $Cx(k) \le c$ $k = 1,2,\ldots,N$

(4.6) $Du(k) \le d$ $k = 0,1,2,\ldots,N-1$

where C and D are $m_1 \times n$ and $m_2 \times p$ respectively, find the
sequences $u(\cdot)$, $x(\cdot)$ that minimize the objective while satisfying
the constraints.

Again $x(\cdot)$ can be explicitly eliminated from the formulation
resulting in a quadratic programming problem of dimension p·N and
requiring a great deal of storage. To exploit the dynamic structure
of the problem we may employ duality theory [10] to write the problem

in the equivalent dual form [38]

(4.7) $\qquad\qquad\qquad$ maximize $\quad \psi(\mu, \eta)$

subject to $\qquad\qquad\qquad \mu \geq 0 \quad, \quad \eta \geq 0$

where the dual function $\quad \psi \quad$ is defined by

$$\psi(\mu, \eta) = \min_{u,x} \sum_{k=1}^{N} \frac{1}{2} x(k)'Q(k)x(k) + \frac{1}{2} u(k-1)'R(k)u(k-1)$$

$$+ \mu(k)'[Cx(k)-c] + \eta(k)'[Du(k-1)-d]$$

with

$$x(k+1) = \Phi x(k) + Bu(k) \quad, \quad k = 0,1,2,\ldots,N-1 ,$$

$$x(0) = x_0 .$$

Assuming for the moment that the dual function can be evaluated efficiently, we note that it is itself a quadratic function of μ, η. Thus the dual problem is a quadratic programming problem but the inequality constraints are simple positivity constraints.

Quadratic programming problems with only positivity constraints can be solved by simple modification of the gradient or conjugate gradient method [38], [39, [40]. Fortunately, the gradient of the dual function is readily computable once the dual function is evaluated. We have

(4.9a) $\qquad \nabla_{\mu(k)} \psi = Cx(k) - c, \quad k = 0,1,2,\ldots,N-1 ,$

(4.9b) $\qquad \nabla_{\eta(k)} \psi = Du(k-1) - d, \quad k = 1,2,\ldots,N ,$

where $x(\cdot)$ and $u(\cdot)$ are the minimizing trajectory and control history determined by evaluation of the dual function. Thus, once an effective procedure for evaluating the dual function is developed, efficient quadratic programming algorithms can be brought to bear on the problem. The result is a nontrivial synthesis of mathematical programming and control theory techniques for the solution of an otherwise almost impossible problem.

Evaluation of the dual function is (for fixed μ and η) a standard quadratic loss control problem with some additional linear terms. It can be shown that its solution can be found recursively from the following relations [38]

$$(4.10a) \quad P(k-1) = Q(k) + \Phi'P(k)S(k) \quad , \quad k = 1,2,\ldots,N \; ,$$

$$(4.10b) \quad P(N) = Q \; ,$$

$$(4.11) \quad M(k) = -[R + BP(k)B]^{-1}BP(k)\Phi \; ,$$

$$(4.12) \quad S(k) = \Phi + BM(k) \; ,$$

$$(4.13) \quad f(k-1) = S'(k)f(k) + M'(k)D'\eta(k) + C'\mu(k-1), \; f(N) = C'\mu(N),$$

$$(4.14) \quad u(k) = M(k+1)x(k) + y(k) \; ,$$

$$(4.15) \quad x(k+1) = \Phi x(k) + Bu(k) \quad , \quad x(0) = x_0 \; ,$$

$$(4.16) \quad y(k) = -[R(k) + B'P(k)B]^{-1}[Bf(k) + B'\eta(k)] \; .$$

Although at first these relations may seem formable we note that (4.10), (4.11), and (4.12) are independent of μ and η and hence $P(k)$, $M(k)$, $S(k)$, $k = 0,1,2,\ldots,N$ can be computed once and used for all future iterations. Calculation of $f(\cdot)$ and $y(\cdot)$ is then accomplished by a single backward recursion and finally $x(\cdot)$ and $u(\cdot)$ are obtained by a forward recursion. The result yields the gradient of the dual problem. Thus, taking advantage of the structure in this manner we have a method for computing the gradient of the dual function that requires one backward and one forward recursion--about the same work as that required for evaluation of the gradient of a primal control problem (in the primal case the forward recursion precedes the backward).

4.3 Speed of Convergence

We conclude by indicating some preliminary results in an area that has to date received no formal attention but which ultimately will certainly play a dominant role in the rational development of

algorithms for control: convergence analysis of algorithms. It can
be argued that convergence analysis, particularly speed of convergence, has unfortunately lagged behind the formal development of
mathematical programming algorithms of all types. The only widely
known result is the analysis of steepest descent. In control theory
the situation is even worse since there has been virtually no analysis even of steepest descent. In what follows we offer an analysis
of a most elementary optimal control problem solved by steepest descent in order to indicate the kind of qualitative conclusions that
can be drawn.

We consider the quadratic problem

$$(4.17) \qquad x(k+1) = \phi x(k) + bu(k) \quad , \quad x(0) = x_0 \; ,$$

$$(4.18) \qquad J = qx(N)^2 + \sum_{k=0}^{N-1} u(k)^2$$

where for simplicity we assume that both $x(k)$ and $u(k)$ are scalars. This problem can be easily solved analytically or numerically
by finding a Riccati-type sequence, but our concern is not with the
question of solving the problem explicitly but with using it as a
model to analyze steepest descent. In expanded form the problem can
be written as a quadratic minimization problem in terms of an N-
dimensional vector U. The quadratic form associated with the expanded problem is

$$(4.19) \qquad \Gamma = I + b^2 q \phi' \phi$$

where ϕ is the row vector

$$\phi = (\phi^{N-1}, \phi^{N-2}, \ldots, 1) \; .$$

Thus, the matrix Γ is equal to the identity plus a diad.

It is well known [10] that for a quadratic problem the method
of steepest descent converges according to

$$(4.20) \qquad J(u_{k+1}) \leq \left(\frac{A-a}{A+a}\right)^2 J(u_k)$$

where   a   and   A   are respectively the smallest and largest eigen-
values of the quadratic form matrix Γ.   Since Γ has a particularly
simple structure for this problem we can see that there is one eigen-
value of magnitude $1 + b^2 q \, \|\Phi\|^2$ and   N-1   eigenvalues of value
unity.   Thus, the rate of convergence can be written explicitly as

$$\left[\frac{b^2 q \, \|\Phi\|^2}{2 + b^2 q \, \|\Phi\|^2} \right]^2 .$$

Several qualitative conclusions can be drawn from this result.
We see that for unstable systems $(|\phi| > 1)$ $\|\Phi\|^2$ grows geometrically
with   N   and hence as   N   increases it becomes increasingly difficult
to solve the problem by steepest descent.   For stable systems
$(|\phi| < 1)$ the difficult increases slowly with   N,   approaching a
finite limit.   In general, the more stable the system the faster
will be the convergence of steepest descent.

Similar convergence analyses can be made for more complex quadratic problems, for continuous as well as discrete systems, and for
algorithms other than steepest descent.   These provide a concrete
basis of comparison among problems and solution techniques.

V.   Conclusions

Although the similarity between mathematical programming and
optimal control has been widely recognized for at least eight years,
interplay between the two fields has been less pronounced than might
be expected, and surely less than potentially possible.   This is
largely due, it seems, to the fact that a major part of the interplay
has been the activity of control theorists rather than mathematical
programmers.   The development of the unified theory of optimization,
for instance, which includes both the continuous-time and discrete-
time optimal control problems as well as classical mathematical programming results, was largely the result of efforts on the part of
control theorists striving to close the gap that existed between
necessary conditions in the two fields.   The majority of existing
algorithms used to solve control problems on a routine basis were
developed from a continuous-time viewpoint by control theorists who

borrowed algorithms from mathematical programming. Mathematical programmers, not motivated by the original practical problems themselves, have, for the most part, taken a fairly passive role in the development of solutions to control problems. There is still much room for interaction, particularly in the areas of convergence analysis and exploitation of structure--areas where mathematical programming has a great tradition.

References

1. Pontryagin, L. S., V. G. Boltyanskii, R. V. Gamkrelidze, and E. F. Mishchenko, _The Mathematical Theory of Optimal Processes_, Interscience, New York, 1962.

2. Canon, Michael D., Clifton D. Cullum, Jr., and Elijah Polak, _Theory of Optimal Control and Mathematical Programming_, Chapters 1,2,4, McGraw-Hill Book Company, New York, 1970.

3. Mangasarian, Olvi L., _Nonlinear Programming_, McGraw-Hill Book Company, New York, 1969.

4. Mangasarian, Olvi L. and S. Fromovitz, "Fritz John Necessary Optimality Conditions in the Presence of Equality and Inequality Constraints," J. Math. Analysis and Applications, Vol. 17, pp. 37-47; 1967.

5. Halkin, Hubert, "A Maximum Principle of the Pontryagin Type for Systems Described by Nonlinear Difference Equations," J. SIAM Control, Vol. 4, No. 1, pp. 90-111; Feb. 1966.

6. Holtzman, J. M., "On the Maximum Principle for Nonlinear Discrete Time Systems," IEEE Trans. for Automatic Control, Vol. 11, pp. 273-274; 1966.

7. Liusternik, L. and V. Sobelev, _Elements of Functional Analysis_, Frederick Ungar, New York, 1961.

8. Hurwicz, L., "Programming in Linear Spaces," _Studies in Linear and Nonlinear Programming_ (K. J. Arrow, L. Hurwicz, and H. Uzawa, eds.), Stanford University Press, Stanford, Calif., pp. 38-102; 1958.

9. Rockafellar, R. T., _Convex Analysis_, Princeton University Press, Princeton, N.J., 1970.

10. Luenberger, David G., _Optimization by Vector Space Methods_, John Wiley and Sons, Inc., New York, 1969.

11. Goldstine, H. H., "A Multiplier Rule in Abstract Spaces," Bull. Amer. Math. Soc., 44, pp. 388-394; 1938.

12. Goldstein, A. A., "Convex Programming and Optimal Control," J. SIAM Control, Vol. 3, No. 1, pp. 142-146; 1965.

13. Halkin, H., "Optimal Control as Programming in Infinite Dimensional Space," in C.I.M.E.: Calculus of Variations, Classical and Modern (E. R. Cremonese, Ed.), Roma, pp. 179-192; 1966.

14. Russell, D. L., "The Kuhn Tucker Conditions in Banach Space with an Application to Control Theory," J. Math. Anal. and Appl., Vol. 15, pp. 200-212; 1966.

15. Varaiya, P. P., "Nonlinear Programming in Banach Spaces," J. SIAM Appl. Math., Vol. 15, No. 2, pp. 284-293; 1967.

16. Gamkrelidze, R. B., "On Some Extremal Problems in the Theory of Differential Equations with Applications to the Theory of Optimal Control," J. SIAM Control, Vol. 3, pp. 106-128; 1965.

17. Neustadt, Lucien W., "An Abstract Variational Theory with Applications to a Broad Class of Optimization Problems. I. General Theory," J. SIAM Control, Vol. 4, No. 3, pp. 505-527; Aug. 1966.

18. -------, "An Abstract Variational Theory with Applications to a Broad Class of Optimization Problems. II. Applications," J. SIAM Control, Vol. 5, No. 1, pp. 90-137; Feb. 1967.

19. -------, "A General Theory of Extremals," J. Computer and System Sciences, Vol. 3, pp. 57-92; 1969.

20. Halkin, H. and L. W. Neustadt, "General Necessary Conditions for Optimization Problems," Proceedings of National Academy of Sciences, Vol. 56, (4), pp. 1066-1071; 1966.

21. Canon, Michael D., Clifton D. Cullum, Jr., and Elijah Polak, "Constrained Minimization Problems in Finite-Dimensional Spaces," J. SIAM Control, Vol. 4, No. 3, pp. 528-547; Aug. 1966.

22. Breakwell, John V., "The Optimization of Trajectories," J. Soc. Indust. Appl. Math., Vol. 7, pp. 215-247; June 1959.

23. Balakrishnan, A. V., "On a New Computing Technique in Optimal Control," J. SIAM Control, Vol. 5; May 1968.

24. Bryson, A. E., Jr. and W. F. Denham, "A Steepest Ascent Method for Optimal Programming Problems," Journal of Applied Mechanics, Vol. 29, No. 2, pp. 247-257; 1962.

25. Kelley, H. J., "Method of Gradients," Optimization Techniques (G. Leitmann, ed.), Academic Press, New York, 1962.

26. Lasdon, L. S., S. K. Mitter, and A. D. Warren, "The Conjugate Gradient Method for Optimal Control Problems," Trans. IEEE, AC-12, pp. 132-138; April 1967.

27. Sinnott, J. F. and D. G. Luenberger, "Solution of Optimal Control Problems by the Method of Conjugate Gradients," reprint of paper presented at JACC, July, 1967.

28. McReynolds, S. R. and A. E. Bryson, Jr., "A Successive Sweep Method for Solving Optimal Control Problems," JACC, pp. 551-555; 1965.

29. Bullock, T. E., "Computation of Optimal Controls by a Method Based on Second Variations," SUDAAR No. 297, Stanford University; Dec. 1966.

30. Bryson, Arthur E., Jr., and Yu-Chi Ho, _Applied Optimal Control_, Blaisdell Publishing Company, Waltham, Mass., 1969.

31. Jacobson, D. H., "Second-Order and Second Variation Methods for Determining Optimal Control: A Comparative Analysis Using Differential Dynamic Programming," Intl. J. of Control, Vol. 7, No. 2, pp. 175-196; 1968.

32. Luenberger, D. G., "A Primal-Dual Method for the Computation of Optimal Control," 2nd International Conference on Computing Methods in Optimization Problems, San Remo, Italy, Sept. 9-13, 1968.

33. Cullum, Jane, "Perturbations of Optimal Control Problems," J. SIAM Control, Vol. 4, No. 3, pp. 473-487; Aug. 1966.

34. Daniel, James W., "The Approximate Minimization of Functionals by Discretization," to appear.

35. Henrici, Peter, _Discrete Variable Methods in Ordinary Differential Equations_, John Wiley and Sons, Inc., New York, 1968.

36. Hamming, R. W., _Numerical Methods for Scientists and Engineers_, McGraw-Hill Book Company, New York, 1962.

37. Dantzig, George B., "Optimal Solution of a Dynamic Leontief Model with Substitution," Econometrica, 23, pp. 295-302; 1955.

38. Maxfield, Robert Roy, "Techniques for Computing Optimal Controls for Linear Systems with Inequality Constraints," Ph.D. Dissertation, Stanford University, Stanford, Calif.; Feb. 1969.

39. Hildreth, C., "A Quadratic Programming Procedure," Naval Research Logistics Quarterly, Vol. 4, pp. 79-85; 1957.

40. Houthakker, H. S., "The Capacity Method of Quadratic Programming," Econometrica, Vol. 28, pp. 62-87; 1960.

III. INTEGER PROGRAMMING

5. INTEGER PROGRAMMING ALGORITHMS: A FRAMEWORK AND STATE-OF-THE-ART SURVEY*†

A. M. GEOFFRION AND R. E. MARSTEN‡

University of California, Los Angeles

A unifying framework is developed to facilitate the understanding of most known computational approaches to integer programming. A number of currently operational algorithms are related to this framework, and prospects for future progress are assessed.

I. Introduction

This paper was written with two complementary purposes in mind. The first is to undertake a survey of existing general purpose integer linear programming algorithms that have been implemented with a notable degree of computational success. The second is to explain a general algorithmic framework for integer programming that we have found quite useful in organizing our understanding of the field.

We have tried to make this paper accessible to as wide an audience as possible by keeping the exposition elementary. Readers wishing to delve more deeply will find suitable references throughout the text. In addition, there are the excellent previous surveys [8], [10], and [9], and also such general frameworks for enumerative type algorithms as [1], [6], [16], [48], [59], and [62].

The proposed general framework is presented in §II, after a discussion of the three underlying concepts on which it is based: separation, relaxation, and fathoming. No claim is made for the originality of these concepts; they appear, either explicitly or implicitly, in many of the references just mentioned. What distinguishes the present treatment is perhaps mainly in the realm of emphasis—especially the central role assigned here to the use of relaxation—and the recognition that one framework serves equally well for algorithms of both enumerative and nonenumerative type.

§III surveys selected current algorithms. In each case it is indicated how the algorithm fits into the general framework, with digressions as necessary to explain any new features of general interest. Computational experience is then cited. The algorithms are grouped according to whether they are based primarily on enumeration (§3.1), Benders Decomposition (§3.2), cutting-planes (§3.3), or on group theory (§3.4). Regrettably, space does not permit discussion of a number of effective algorithms proposed recently for special applications such as vessel scheduling [3], facilities location [23], and crew scheduling [52]. A review of numerous specialized algorithms can be found in [9].

§IV attempts to summarize the current state-of-the-art for integer programming, and presents some opinions regarding promising directions for the future development of the field.

Throughout this paper, the canonical statement of the mixed integer linear programming problem is taken to be

$$\text{(MIP)} \qquad \text{Minimize}_{x \geq 0; \, y \geq 0} \quad cx + dy$$

$$\text{subject to} \quad Cx + Dy \leq b, \qquad x \text{ integer,}$$

* Received March 1971.

† This research was supported by the National Science Foundation under Grant GP-26294. An earlier version was presented at the Third Operations Research Symposium, jointly sponsored by CPPA and TAPPI, Montreal, Quebec, August 24–26, 1970.

‡ Now at Northwestern University.

where C and D are matrices and c, d, b, x and y are vectors of appropriate dimensions. By taking d and D to be null, this problem specializes to the "pure integer" linear program. It will be assumed for simplicity that the feasible region of (MIP) is bounded if it is not empty.

II. A General Framework

The general algorithmic framework is built upon three key notions: separation, relaxation, and fathoming. Each of these is discussed in turn before the general framework itself is presented.

2.1. *Separation*

For any optimization problem (P), let $F(P)$ denote its set of feasible solutions. Problem (P) is said to be *separated* into subproblems (P_1), $\cdots$, (P_q) if the following conditions hold:

(S1) Every feasible solution of (P) is a feasible solution of exactly one of the subproblems (P_1), $\cdots$, (P_q).

(S2) A feasible solution of any of the subproblems (P_1), $\cdots$, (P_q) is a feasible solution of (P).

These conditions assert that $F(P_1)$, $\cdots$, $F(P_q)$ is a partition of $F(P)$. The subproblems (P_1), $\cdots$, (P_q) are called *descendants* of (P). Creating descendants of the descendants of (P) is equivalent to refining the partition of $F(P)$.

Our interest in separation is that it enables an obvious divide-and-conquer strategy for solving (P). Leaving aside for a moment the important question of *how* one separates a problem that is difficult to solve, we can sketch a rudimentary strategy of this type as follows. First make a reasonable effort to solve (P). If this effort is unsuccessful, separate (P) into two or more subproblems, thereby initiating what will be called a *candidate list* of subproblems. Extract one of the subproblems from this list—call it the current *candidate problem* (CP)—and attempt to solve it. If it can be solved with a reasonable amount of effort, go back to the candidate list and extract a new candidate problem to be attempted; otherwise, separate (CP) and add its descendants to the candidate list. Continue in this fashion until the candidate list is exhausted. If we refer to the best solution found so far to any candidate problem as the current *incumbent*, then the final incumbent must obviously be an optimal solution of (P) (if all candidate problems were infeasible, then so is (P)).

The finite termination of such an approach is obviously assured if $F(P)$ is finite [provided that there is a finite limit on the number of times a degenerate separation of a candidate problem can occur due to all but one descendant being infeasible]. The usefulness of this strategy depends, of course, upon a sufficient level of success at solving the candidate problems without further separation. Considerable subsequent discussion will be addressed to this point.

The most popular way of separating an integer programming problem is by means of contradictory constraints on a single integer variable (the separation or branching variable). For example, if x_4 is declared to be a binary variable, then (MIP) can be separated into two subproblems by means of the mutually exclusive and exhaustive constraints "$x_4 = 0$" and "$x_4 = 1$". For general integer variables there are other possibilities. For example, if x_5 is declared to be an integer between 0 and 4, then the constraints $x_5 = 0$, $x_5 = 1$, $x_5 = 2$, $x_5 = 3$, $x_5 = 4$ are mutually exclusive and exhaustive, as are $0 \leq x_5 \leq 2$, $3 \leq x_5 \leq 4$. Several possible rules for choosing the integer variable on which the separation should be based will be mentioned in §III.

Sometimes special problem structure can be exploited by devising more powerful separation techniques. Beale and Tomlin [12] report success along these lines for problems containing "multiple choice" constraints such as $x_1 + x_2 + x_3 + x_4 + x_5 = 1$, where each of the variables is binary. A separation can be effected here by mutually exclusive conditions of the sort $x_1 + x_2 + x_3 = 0$, $x_4 + x_5 = 0$. In this case the freedom of several variables is restricted simultaneously.

2.2. *Relaxation*

Any constrained optimization problem (P) can be *relaxed* by loosening its constraints, resulting in a new problem (P_R). For example, one may simply omit some of the constraints of (P). Other possibilities for relaxation include dropping the integrality conditions or the nonnegativity conditions on the variables of (P). The only requirement for (P_R) to be a valid relaxation of (P) is: $F(P) \subseteq F(P_R)$. This defining property of relaxation trivially implies the following relationships, where, by convention, (P) is taken to be a minimization problem:

(R1) If (P_R) has no feasible solutions, then the same is true of (P).

(R2) The minimal value of (P) is no less than the minimal value of (P_R).

(R3) If an optimal solution of (P_R) is feasible in (P), then it is an optimal solution of (P).

In selecting between alternative types of relaxation for a given problem, there are two main criteria to be considered. On the one hand, it is desirable for the relaxed problem to be significantly easier to solve than the original. On the other hand, for reasons that will become clear, one would like (P_R) to yield an optimal solution of (P) via (R3) or, failing that, the minimal value of (P_R) should be as close as possible to that of (P). Unfortunately these objectives tend to be antagonistic. In general, the easier (P_R) is to solve, the greater is the "gap" between the original and relaxed problems.

In §III the principal relaxation techniques used in integer programming will be illustrated by reference to various algorithms. By far the most popular type of relaxation for an integer linear program is to drop all integrality requirements on the variables. The resulting ordinary linear program is often an excellent compromise between the two criteria mentioned above. A number of other interesting types of relaxation can be gleaned, for instance, from [40], [42], [60], [75], [76]. A discussion of the role of relaxation in continuous mathematical programming can be found in [30, §3.3].

2.3. *The Fathoming Criteria*

In §2.1 a simple divide-and-conquer strategy based on the notion of separation was briefly sketched. A sequence of candidate problems must be examined, and whenever one of them cannot be solved with a reasonable amount of effort, it must be separated and its descendants must be examined subsequently. Use of the vague phrase "reasonable amount of effort" was deliberate, for the success of the strategy depends to a great extent on the judicious choice of how and how hard to try to solve each candidate problem. It is in this regard that the concept of relaxation plays an important role; rather than trying to deal with the candidate problem itself, one deals instead with some relaxation of it. The so-called fathoming criteria introduced below are an attempt to clarify and formalize this role.

Let (CP) be a typical candidate problem arising from the attempt to solve (P). The ultimate objective in dealing with (CP) is to determine whether its feasible region

$F(CP)$ may possibly contain an optimal solution of (P) and, if so, to find it. If it can be ascertained by some means that $F(CP)$ cannot contain a feasible solution better than the incumbent (the best feasible solution yet found), this is certainly good enough to dismiss (CP) from further consideration, and we say that (CP) has been *fathomed*. Or if an optimal solution of (CP) can actually be found, we also say that (CP) has been fathomed. In either case, the candidate problem has been entirely accounted for and can be discarded without further separation.

It is useful to distinguish three general types of fathoming, all of them based on relaxation. We shall suppose that a particular relaxation (CP_R) of the candidate (CP) has been decided upon and that (CP_R) has been solved. All problems are to minimize, $v(\cdot)$ denotes the optimal value of problem $(\cdot)$, and Z^* is the value of the incumbent (let $Z^* = +\infty$ if no feasible solution of (P) has yet been found).

If (CP_R) has no feasible solution, then by (R1) the same is true of (CP). Thus $F(CP)$ is empty and cannot possibly contain an optimal solution of (P). In this event (CP) is certainly fathomed.

Secondly, if

$$(1) \qquad\qquad v(CP) \geqq Z^*$$

then $F(CP)$ cannot possibly contain a feasible solution of (P) better than the incumbent. Unfortunately, $v(CP)$ is unknown, but $v(CP_R)$ *is* known and by (R2) we have $v(CP) \geqq v(CP_R)$. It follows that (CP) has been fathomed if

$$(2) \qquad\qquad v(CP_R) \geqq Z^*.$$

Note that it is possible to have $v(CP) \geqq Z^* > v(CP_R)$, so that condition (1) is satisfied but condition (2) is not satisfied. In this case the outcome of (CP_R) will not permit discarding (CP). It now becomes evident why, as mentioned earlier, it is desirable to have $v(CP_R)$ as close to $v(CP)$ as possible.

Thirdly, suppose that an optimal solution of (CP_R) has been found which happens to be feasible in (CP). Then by (R3) this solution must be optimal for (CP) and hence (CP) is fathomed. By (S2), of course, it is also feasible in (P) and so becomes the new incumbent if its value is less than Z^*.

The above discussion can be summarized and slightly generalized in terms of three *fathoming criteria*. Candidate problem (CP) is fathomed if any one of these criteria is satisfied.

(FC1) An analysis of (CP_R) reveals that (CP) has no feasible solution; e.g., $F(CP_R) = \varnothing$.

(FC2) An analysis of (CP_R) reveals that (CP) has no feasible solution better than the incumbent; e.g., $v(CP_R) \geqq Z^*$.

(FC3) An analysis of (CP_R) reveals an optimal solution of (CP); e.g., an optimal solution of (CP_R) is found which happens to be feasible in (CP).

Among the various algorithms that have been proposed, there is considerable variation in the kinds of "analysis" actually employed to implement these criteria. The standard analysis, which conforms to the discussion above, is to solve (CP_R) optimally and then check whether $F(CP_R) = \varnothing$ or $v(CP_R) \geqq Z^*$ or the optimal solution obtained for (CP_R) is feasible in (CP). But there are deviations from this norm in both directions. Some algorithms do not go so far as to solve (CP_R) optimally, but merely employ sufficient conditions for the tests $F(CP_R) = \varnothing$ and $v(CP_R) \geqq Z^*$ (e.g., any lower bound $\hat{v}(CP_R)$ on $v(CP_R)$—say a good suboptimal solution of the dual problem if (CP_R) is a linear program—leads to the sufficient condition $\hat{v}(CP_R) \geqq Z^*$). Other algorithms not only go so far as to solve (CP_R) optimally, but continue the analysis even further in an effort to account at least partially for whatever "gap" there may be between (CP) and (CP_R). As we shall see in §III, when (CP_R) is a linear program this further analysis is often a kind of postoptimality study aimed at finding a better lower bound $\hat{v}(CP)$ on the value of (CP) than $v(CP_R)$ itself; the test $\hat{v}(CP) \geqq Z^*$ is then stronger than the standard test $v(CP_R) \geqq Z^*$.

There are two alternative courses of action that may be taken if a particular relaxation (CP_R) of (CP) does not pass any of the fathoming criteria. The first is immediately to separate (CP) and add its descendants to the candidate list. The second alternative is to persist in trying to fathom (CP), as by choosing a new relaxation (CP_R'). For example, some previously ignored constraints can be reimposed. Each new relaxation provides another chance to pass one of the fathoming criteria. The relaxation of (CP) can be modified as many times as desired, but if it appears that too much time is being expended on this one candidate problem, a switch ought to be made to the first alternative—separation.

2.4. *General Procedure*

The ingredients of a general procedure for solving (MIP) are now at hand. A flow-chart is given in Figure 1.

Step 1. Initialize the candidate list to consist of (MIP) alone, and Z^* to be an arbitrarily large number.

Step 2. Stop if the candidate list is empty: if there exists an incumbent then it must be optimal in (MIP), otherwise (MIP) has no feasible solution.

Step 3. Select one of the problems from the candidate list to become the current candidate problem (CP).

Step 4. Choose a relaxation (CP_R) of (CP).

Step 5. Apply an appropriate algorithm to (CP_R).

Step 6. Fathoming Criterion 1. If the outcome of Step 5 reveals (CP) to be infeasible (e.g., $F(CP_R) = \varnothing$), go to Step 2.

Step 7. Fathoming Criterion 2. If the outcome of Step 5 reveals that (CP) has no feasible solution better than the incumbent (e.g., $v(CP_R) \geqq Z^*$), go to Step 2.

Step 8. Fathoming Criterion 3. If the outcome of Step 5 reveals an optimal solution of (CP) [e.g., an optimal solution of (CP_R) is feasible in (CP)], go to Step 12.

Step 9. Decide whether or not to persist in attempting to fathom (CP). If so, go to Step 10; otherwise go to Step 11.

Step 10. Modify (perhaps by tightening) the relaxation of (CP) and return to Step 5.

Step 11. Separate (CP) and add its descendants to the candidate list. Go to Step 2.

Step 12. A feasible solution of (MIP) has been found; if $v(CP) < Z^*$, record this solution as the new incumbent and set $Z^* = v(CP)$. Go to Step 2.

Remarks. Step 1. Prior knowledge about (MIP) can be used to improve the (default) initializations given above for Step 1. One may know an upper bound Z_0 on the optimal value of (MIP), for example, in which case initializing Z^* to the value Z_0 will cause the procedure to look only for solutions with value less than Z_0. One may also have reason to prefer starting with an initial candidate list obtained by making certain prior separations of (MIP). This too is permissible, and simply leads to an initial candidate list with more than one member.

Step 2. If Z^* was initialized at Z_0 rather than an arbitrarily large number, the absence of an incumbent upon termination should be interpreted as proof that (MIP) has no feasible solution with value less than Z_0.

Step 3. Different rules for selecting the next candidate problem lead to different

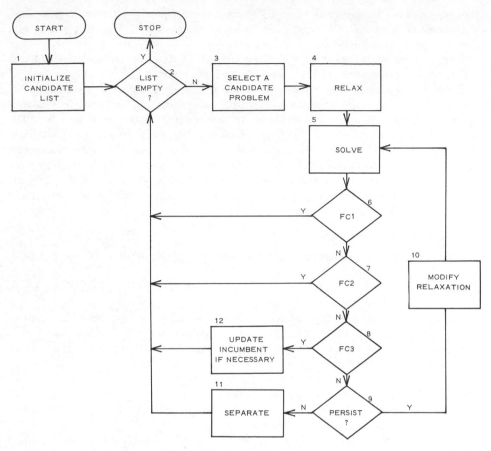

FIGURE 1. A general procedure.

patterns of enumeration. Two main types of rules have been employed, as will be illustrated in the next part of this paper: Last-In-First-Out (*LIFO*) and *Priority*. Under LIFO, the problem selected is always the last one that was added to the candidate list. The two main advantages are that the bookkeeping associated with maintaining the candidate list can be done very compactly, and reoptimization at Step 5 is greatly facilitated (see the remark on Step 5 below). The main disadvantage is that there is less flexibility to control the enumerative process; occasions do arise when some older problem in the candidate list may look more promising. Priority rules, on the other hand, achieve flexibility by selecting the problem with the best priority index, where the index for each problem may be given by (or be computed from) a tag affixed at the time of creation. The tag may be, for instance, a lower bound on the optimal value. Unfortunately, such flexibility must be purchased at the price of more cumbersome bookkeeping and lessened opportunity to reoptimize efficiently at Step 5. The most effective rules for Step 3 therefore strike a balance between the LIFO and Priority extremes.

Step 4. It is conceivable that (CP_R) is so relaxed by comparison with (CP) that it has unbounded optimal value. Since the feasible region of (MIP) was assumed bounded, however, we may guarantee without loss of generality that (CP_R) is likewise bounded by including adequately large bounds on the variables of (CP_R) if necessary.

Step 5. It is important not to solve each candidate problem from scratch, but rather to update earlier solutions efficiently whenever possible. Whether this is practical depends on the type of algorithm used for this step and on the choices made at Steps 3, 4, and 11. Note that it may not always be necessary to solve (CP_R) completely, since the sufficient conditions employed to implement Steps 6, 7 and 8 may be designed to operate with less information than an optimal solution of (CP_R).

Steps 6, 7, and 8. It bears emphasis that one need only apply convenient *sufficient* conditions at these steps, for to fathom with absolute certainty could be computationally onerous. The order of these steps is not at all rigid. In fact, it is best to conceptualize Steps 5 through 8 as a single unit devoted to the attempted fathoming of (CP).

Step 9. A very simple rule would be: give up after a certain number of unsuccessful attempts to fathom (CP). A more sophisticated stopping rule could be based on the apparent rate of progress toward an optimal solution of (CP).

Step 12. An additional function can be performed if an improved feasible solution of (MIP) has been found and the problems in the candidate list are tagged with lower bounds on their optimal values: purge the list of those problems whose lower bounds are not smaller than the new Z^*, for they obviously cannot contain a feasible solution better than the new incumbent.

This general procedure admits a great deal of flexibility, as will be evident from the diversity of algorithms discussed in the next section. It should also be clear that, although the scope of this paper is confined to integer linear programs, the framework given here extends readily to integer nonlinear programs and many other combinatorial optimization problems that do not have a natural or efficient mathematical programming formulation.

III. Current Algorithms

We now consider a number of recent general purpose algorithms whose computational effectiveness has been empirically demonstrated for medium-to-large problems. Each algorithm is described in terms of the general framework, with digressions as necessary to explain novel features of general interest. Computational experience is cited, most of which is as yet unpublished.

First a number of enumerative (implicit enumeration, branch-and-bound) algorithms are covered, including those of [22], [61] (OPHELIE MIXED), [14] (MPSX-MIP), [69], [70] (UMPIRE), [29] (RIP30C), [21], [58], and [43] (MARISABETH). Benders Decomposition is explained in the next section, and computational results from [2] (MIDAS-2), [19] (FMPS-MIP), and [51] (IPE) are reported. In §3.3 the cutting-plane approach is briefly discussed, including computational results from [55], [56]. We treat the group theory approach in a manner strongly influenced by [39] (IPA).

Any omissions of computationally successful general purpose codes should be interpreted as due to lack of information on our part, rather than as the exercise of personal opinion.

The reader is cautioned to interpret the numerical results quoted here with care, especially when it comes to making effectiveness comparisons between different algorithms. The figures quoted here are believed accurate, but the description of the problems run and the manner in which each code was actually operated are necessarily incomplete. Taken in the aggregate, however, the results reported do give a fairly good feeling for the current state-of-the-art.

3.1. *Enumerative Algorithms*

By "enumerative" algorithms we mean to include all those which come under the popular headings of "implicit enumeration" and "branch-and-bound." Such algorithms methodically search the set of possible integer solutions in such a way that not all possibilities need be considered individually.

It seems natural to classify enumerative algorithms into two categories: those which base their fathoming tests primarily on the logical implications of the problem constraints, and those which base them primarily on associated linear programs (derived by ignoring integrality requirements). The first category of algorithms, applicable only to all-integer problems, was stimulated in this country to a great extent by Balas [4]. Work in this vein includes [28], [33], [41], [49], [74], and many more. Although there have been some notable successes in solving problems of very special structure (e.g., [26], [44], [60]), it appears that in general computing times tend to increase approximately exponentially with the number of variables. This is undoubtedly the reason why, so far as we know, no computer code for an algorithm of this type has yet proven to be of *general* practical utility.

The other category, however, has spawned a number of computer codes of general utility for mixed as well as pure integer programs. Work employing linear programming as the primary fathoming device was stimulated by the seminal contribution of Land and Doig [46]. Before discussing several recent developments in this vein, it will be useful to review the generally superior variant of the Land and Doig algorithm proposed by Dakin [20].

The following summary indicates how Dakin's algorithm for (MIP) fits into the general framework.

Algorithm of Dakin.

Step 1. Standard.
Step 2. Standard.
Step 3. LIFO rule.
Step 4. Relax all integrality requirements.
Step 5. Solve by a linear programming algorithm.
Step 6. Standard.[1]
Step 7. Standard.[1]
Step 8. Standard.[1]
Step 9. Always go to Step 11.
Step 10. Omit.
Step 11. Dichotomize the current candidate problem via the alternative constraints $x_j \leq [\bar{x}_j]$ and $x_j \geq [\bar{x}_j] + 1$, where $\bar{x}_j$ is a fractional-valued x-variable in the solution found at Step 5.[2] See the text for the particular rule used to select the separation variable and the order in which the new candidate problems are added to the list.
Step 12. Standard.

Step 11 is one of the main respects in which the various enumerative algorithms differ. Dakin's interval dichotomization constraints have been widely adopted, but many other ways have been suggested for selecting the variable on which to base the

---

[1] The "standard" option for this step is always the parenthetical one given in §2.4.
[2] $[x]$ stands for the integer part of x.

separation, and the order in which to add the new candidate problems to the list (when this matters). Dakin selected the separation variable by first estimating, for each alternative constraint of each fractional-valued variable, the amount by which the optimal value of the current linear program would increase if the constraint were introduced. The estimate used was the increase that would actually occur during the very first dual simplex method iteration, assuming that this method is used for reoptimization. The separation variable selected is the one corresponding to the largest such estimate, with the corresponding new candidate problem being placed on the list last (after its alternative).

For computational experience with this basic algorithm, see the original paper by Dakin, [15], and [73].

3.1.1. *Davis, Kendrick and Weitzman* [22] *and Penalties.* The "estimates" used at Step 11 by Dakin contain the essence of an important development commonly known as *penalties*. Widespread appreciation of the significance of penalties followed the publication of [24], which gave an independent development in the context of a quite rudimentary enumerative scheme. We prefer to describe here the work of Davis, Kendrick and Weitzman (DKW), however, since they gave one of the first systematic developments of penalties in the context of a powerful enumerative scheme.

The DKW paper addresses mixed integer programs in which all integer variables are binary. In terms of the general framework, its outline would be the same as that given above for [20] except for Steps 3, 7 and 11. Leaving aside for a moment the details of computing and using penalties, these three steps can be described in general terms as follows.

Step 3. Select the candidate problem with the smallest associated bound (see Step 11).[3]

Step 7. If $v(CP_R) + PEN(CP_R) \geq Z^*$, go to Step 2.

Step 11. Dichotomize (CP) in the manner of Dakin. The variable selected is the one with the greatest "up penalty" or "down penalty"; the associated bounds of the two descendants are $v(CP_R)$ plus the appropriate penalties.

Encouraging computational results have been obtained for a number of test problems. The largest of these, having to do with hospital scheduling, had 197 constraints, 217 continuous variables, and 100 0-1 variables; it was solved in $5\frac{1}{2}$ minutes on an IBM 7094. More recently, it is reported [27] that an investment planning model for the electric power industry with 290 constraints, 410 continuous variables, and 92 binary variables was solved in 21 minutes on an IBM 360/85.

It remains now to discuss the penalties referred to above. The key fact is that an optimal linear programming tableau contains information that enables an estimate to be readily constructed of the amount by which the optimal value would change if any one of the variables were forced to assume a particular integer value. The situation can be visualized in terms of the following figure, which depicts the optimal value of a minimizing linear program—say the one solved at Step 5—as a function, $V_j(x_j)$, of the value of x_j considered as a parameter. It is well known and not difficult to show that $V_j(x_j)$ is a convex piecewise-linear function attaining its minimum at $\bar{x}_j$, the optimal value of x_j in the linear program. (Although this possibility is not illustrated, it is possible for the linear program to become infeasible for x_j too large

---

[3] DKW also suggested the following modification: if no feasible integer solution has yet been found and separation has just taken place (rather than fathoming), limit the selection to the two problems just created (choose the one with the smallest associated bound).

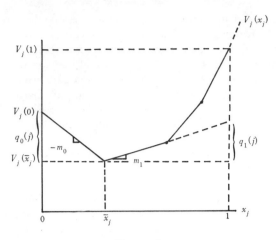

FIGURE 2.

or too small; in this case $V_j(x_j)$ could be considered as having infinite value.) Designate by m_1 ($-m_0$) the right-hand (left-hand) derivative or slope of V_j at the point $\bar{x}_j$. These derivatives can usually be computed directly from the optimal tableau, and enable estimation of $V_j(0)$ and $V_j(1)$ by linear extrapolation. The resulting estimate of $V_j(0)$ is $V_j(\bar{x}_j) + m_0\bar{x}_j$, which happens to equal the true value in this case. The estimate of $V_j(1)$ is $V_j(\bar{x}_j) + m_1(1 - \bar{x}_j)$, which undershoots the true value. In general, by convexity these estimates can never exceed the true values. The quantity $m_0\bar{x}_j$ will be denoted by $q_0(j)$ and called the *simple down penalty* associated with x_j; $m_1(1 - \bar{x}_j)$ will be denoted by $q_1(j)$, the *simple up penalty*.

Clearly $q_u(j) \geqq 0$ and $V_j(u) \geqq V_j(\bar{x}_j) + q_u(j)$ for all j and $u = 0$ or 1. Furthermore, since V_j must increase by at least Min $\{q_0(j), q_1(j)\}$ if x_j is to be an integer, the optimal value of the linear program must increase by at least

$$PEN \triangleq \text{Max}_j \{\text{Min} \{q_0(j), q_1(j)\}\} \geqq 0$$

for any solution that is all integer.

This analysis directly yields the improved version of Step 7 given above for DKW, since

$$v(CP) \geqq v(CP_R) + PEN(CP_R) \geqq v(CP_R)$$

(the new notation makes explicit the dependence of PEN on (CP_R)).

Before discussing the use of penalties at Step 11, we mention one rather obvious further use of penalties at Step 7. Even if $v(CP_R) + PEN(CP_R) < Z^*$, it is still possible to have $v(CP_R) + q_{\hat{u}}(\hat{j}) \geqq Z^*$ for some $\hat{j}$ and $\hat{u} = 0$ or 1. In this case the variable x_j can be fixed at the value $1 - \hat{u}$ in the current candidate problem and in any of its descendants.

At Step 11 one computes Max$_j$ {Max $\{q_0(j), q_1(j)\}\}$. If $q_{\hat{u}}(r)$ achieves this maximum, then x_r is chosen to be the separation variable. A lower bound on the optimal value of the descendant associated with $x_r = \hat{u}$ is $v(CP_R) + q_{\hat{u}}(r)$, while one associated with the other descendant is $v(CP_R) + PEN(CP_R)$ [note that while

$$v(CP_R) + q_{1-\hat{u}}(r)$$

TABLE 1

Computational Results with OPHELIE MIXED

Binary Var.	12	99	114	150	20	48	54	93	117	50	55	81	69	240
Contin. Var.	4,146	839	0	611	2,595	947	252	20	1,970	0	284	1,074	4,858	102
Rows	2,042	1,065	60	566	423	457	279	63	2,164	6	205	264	1,574	397
Nonzero Entries	31,137	5,220	664	2,867	9,088	8,148	774		13,860				30,438	2,616
Application					WH LOC.	RFNRY.	DISTR.		INV.PLN.			PROD.	DISTR.	CHEM.INV.
Init. LP (Min.)														
CP	660.	1.4	0.01	0.4					t					t
IO	1,800.	3.3	0.07	1.2									141.7	
SS														
Total Time (Min.)														
CP	735.	2.6	0.3	4.0	7.0			4.0	23.2+t	0.3	3–4		901.7	4.6+t
IO	2,010.	9.6	2.0	19.3	0			25.0		0.8	0			
SS	0	0	0	0	0	41.8	3.3	15.5		0	0	3.3	0	0
Solution Tol. %	0	0	0	0		11%	0	0	0.7%	0	0	<1%	0	0
Reference	⟵——— Edney [25] ———⟶				⟵——— Blondeau [17] ———⟶					⟵——— Sommer [67] ———⟶				

is also a valid lower bound, the given bound is superior since

$$q_{1-\hat{u}}(r) = \text{Min } \{q_0(r), q_1(r)\} \leqq PEN(CP_R)].$$

3.1.2. *OPHELIE MIXED* (*Roy, Benayoun and Tergny* [61]). The algorithm described in [61] is quite close to the DKW algorithm just described. The only discernible differences are in the elaborate heuristics used at Steps 3 and 11.

Step 3 mainly uses the priority rule of selecting the candidate problem with the most promising bound, but secondary criteria are also employed to take into consideration the probable ease with which a given candidate problem could be reoptimized and the plausibility of the associated bounds (e.g., the more variables fixed the better the likely quality of the bound).

Step 11 is done in the manner of DKW, except that the choice of separation variable may be limited to a set determined by a "hierarchy graph" reflecting the analyst's opinions concerning the relative importance of different variables.

The algorithm has been commercially implemented for the CDC 6600 as OPHELIE MIXED by SEMA using the OPHELIE II linear programming system for Step 5. Table 1 quotes some unpublished results obtained from private sources. In interpreting the computing times, all on the CDC 6600, it should be kept in mind that customer billing is done on the basis of "system seconds" (SS), which is central processing (CP) seconds plus {IO seconds × fraction of central memory used}. System seconds is therefore the best measure of efficiency. See also the extensive published computational results in [61].

3.1.3. *MPSX-MIP* (*Bénichou et al.* [14]). The algorithm employed in [14] coincides with that of Dakin in its essentials except for Steps 3 and 11. Step 3 employs a compromise LIFO/Priority strategy, while Step 11 makes use of a preconceived priority order for selecting the separation variable. More specifically, the version of these steps used in most of the computations is as follows.

Step 3. Select a problem from among those designated "Class 2" if possible; otherwise, from among those designated "Class 1"; if both of these classes are empty, select from the remaining unclassified problems. When selecting from Class 2, the choice is arbitrary; when from Class 1, select the problem with the best "estimation" (see below), redesignate both this problem and its alternative (the one created at the same execution of Step 11—it has the same estimation) as Class 2, and cancel any remaining Class 1 designations; in selecting from the problems with no class designation, choose the problem with the best estimation and designate both this problem and its alternative as Class 2.

Step 11. Dichotomize (CP) in the manner of Dakin. Choose the separation variable from among those with a fractional part of between 0.1 and 0.9 unless this subset is empty. In either case, select according to a priority order which ranks variables in decreasing order of their absolute cost coefficients. Designate the two descendants as Class 1, and associate with each the estimation associated with the candidate problem being dichotomized.

The "estimation" referred to is a heuristic formula that estimates $v(CP)$ by

$$v(CP_R) + \sum_j \text{Min } \{PCL_j(\bar{x}_j - [\bar{x}_j]), PCU_j([\bar{x}_j] + 1 - \bar{x}_j)\},$$

where the sum is over the fractional-valued variables and PCL_j (PCU_j) is a lower (upper) "pseudocost" for variable j. Pseudocosts are quite similar in spirit to penalties, but are empirically determined rather than being derived from the optimal tableau like $q_u(j)$. It appears that they are taken to be the most recently observed *actual* rate of increase in the optimal value of a relaxed candidate problem due to forcing a fractional variable to 0 or 1.

Bénichou et al. have suggested a number of variants and embellishments on the above version of their algorithm, such as using criteria other than "estimations" at Step 3 and alternative priority orders at Step 11. Space does not permit a detailed description here.

A commercial implementation has been carried out by IBM France based on IBM's Extended Mathematical Programming System. A number of quite large problems have been solved. One production scheduling problem with 721 constraints, 39 binary variables and 1117 continuous variables was solved in 18.3 minutes on a 360/65 once the first LP solution was obtained. Another with 157 constraints and 78 binary variables was solved in 6.07 minutes (not including first LP) on an IBM 370/155. The largest problem reported in terms of integer variables had 590 of them with 69 constraints, and was solved in 25.7 minutes (not including first LP) on a 360/65.

3.1.4. *UMPIRE and Improved Penalty Bounds (Tomlin, [69], [70]).* Tomlin [69] (see also [70]) proposed two ways of improving the simple penalty bounds discussed previously in §3.1.1, and has tested these improvements in an implementation of the algorithm of Beale and Small [11] for general mixed integer programs. The latter algorithm will be reviewed below.

Tomlin's first method for obtaining improved bounds is based on the following simple observation: forcing a fractional variable in a linear programming solution to an integer value requires increasing at least one nonbasic variable above zero, but any such increase must be by *at least unity* if this nonbasic variable is specified as integer in the original problem statement. The latter condition is ignored in the earlier derivation of simple penalties. Tomlin showed how this requirement could be used to yield stronger up and down penalty bounds.

The smaller of the new up and down penalties associated with a given fractional variable is a lower bound on the increase in optimal value caused by forcing that variable to be integer (cf., Min $\{q_0(j), q_1(j)\}$ in §3.1.1). Tomlin has shown, however, that a Gomory cut associated with the same fractional variable yields a still better bound than the smaller of the improved up and down penalties. Such a cut is generated in the usual way, and a linear estimate of the increase in optimal value caused by adding it to the linear program is constructed from information available in the final tableau in the same manner by which one obtains $q_u(j)$. The resulting bound might be called the Gomory-penalty $P_G(j)$ associated with the given fractional variable (*cf.*, p. 907 of [49]).

These improved penalties can be used in the obvious way to improve any algorithm that employs simple penalties. Tomlin chose to incorporate them in the algorithm described in [11], which is the same as that of Dakin described previously except for Steps 7 and 11. These two steps happen also to be the ones where penalties are used. They become as follows when Tomlin's more powerful penalties are used in place of the simple penalties employed by Beale and Small.

Step 7. If $v(CP_R) + \text{Max}_j \{P_G(j)\} \geq Z^*$, go to Step 2.

Step 11. Dichotomize (CP) in the manner of Dakin. Choose for the separation variable the one with the largest improved up or down penalty, put the corresponding descendant next on the candidate list, and finally place its alternative at the very end of the list.

It is reported in [69] that using the more powerful penalties in place of simple penalties can reduce the work to solve all-integer or predominantly integer problems by

50 % or more. For a capital budgeting problem with 50 binary variables and 5 constraints, for example, the solution time on a Univac 1108 was reduced from 94 sec. to 44 sec.

Additional computational experience with these improved penalties is cited in [70]. The algorithm used was essentially the same as the one above except that a different type of separation was employed at Step 11 for variables appearing in multiple-choice type constraints (i.e., constraints of the form $\sum_{j \in J} x_j = 1$). The type of separation used for these variables, first described in [12], is of the form: $x_j = 0$, all $j \in J_1$ or $x_j = 0$, all $j \in J - J_1$, where J_1 is some subset of J. Penalties appropriate to this type of dichotomization are derived. The implementation, within the UMPIRE system, was done by Scientific Control Systems Ltd. for the Univac 1108. Computational experience with a number of quite large problems is reported. One had about 250 rows, 200 continuous variables and 164 binary variables; another had about 200 rows, 700 continuous variables and 40 binary variables; and two others had about 1000 constraints and 200 binary variables. Although a strict optimal solution was not obtained for any of these problems, satisfactory integer solutions are claimed. At the very least, it can be concluded that the modifications made to the original Beale and Small algorithm—improved penalties and a special facility for handling multiple-choice variables—greatly improve the performance.

3.1.5. *RIP30C and Surrogate Constraints (Geoffrion* [29]). All of the above algorithms use linear programming as the main fathoming device, and make essentially no use of the logical fathoming devices employed by the additive algorithm of [4] and others of this school mentioned in the introduction to §3.1. The main limitation of these fathoming devices is that they apply at reasonable computational cost to only one constraint at a time. The joint logical implications of several constraints taken simultaneously are therefore lost, with the apparent result that computation times tend to increase approximately exponentially with the number of integer variables for methods relying only on logical fathoming devices. In [29] it is shown how a "surrogate" constraint (see Glover [33]) can be constructed which captures more of the joint logical implications of the entire set of original constraints. It was demonstrated that the dual variables in the solution of a candidate problem as a linear program yield a surrogate constraint that is as "strong" as possible in an appropriate sense. This constraint is constructed by taking a nonnegative linear combination of the original constraints using the values of the dual variables as weights, and then adding to it the constraint that the objective function must have a better value than that of the incumbent. A closely related construct was developed independently in [5].

A computer code (RIP30C) was built to determine how much improvement, if any, such strongest surrogate constraints would make in Balas' additive algorithm as organized in [28]. The search strategy of that algorithm is LIFO at Step 3 and uses a "least total infeasibility after branching" rule at Step 11 to select the separation variable. The standard logical fathoming strategy for Steps 4–10, which will not be reviewed here, was supplemented with a linear programming subroutine that could be used optionally to generate strongest surrogate constraints. Tests on a wide variety of all 0-1 problems showed decisively that the modification leads to a great improvement in performance. Of even more significance than the large observed reductions in computing times for specific problems, perhaps, was the observed behavior of computing times as a function of problem size. Sequences were run of set-covering, optimal routing, and capital budgeting problems of varying sizes up to 90 0-1 variables. The modification appeared to mitigate solution-time dependence on the number of variables from an exponential to a low-order polynomial increase. The dependence was approximately linear for the first two problem classes, with 90-variable problems typically being solved in about 15 seconds on an IBM 7044; and approximately cubic for the third class, with 80-variable problems typically solved in under 2 minutes.

The point is sometimes raised that since linear programming is used to generate strongest surrogate constraints, it is possible to attribute the improved efficiency to the fathoming power of the linear program itself rather than to the power of surrogate constraints. Some light can be thrown on this issue by running the RIP30C code with the linear programming routine fully operative but with no memory of surrogate constraints from one candidate problem to the next. For a representative battery of 8 test problems in the range of 44 to 90 variables, this increased the running time per problem by an average of 100 % (the median was 42 %). This is if anything a conservative rather than true indication of the value of surrogate constraints, however, because the standard LP fathoming tests at Steps 6 and 7—which remained operational—are really equivalent to the construction of a binary infeasible surrogate constraint, a task that could often be done by means other than by linear programming. On the other hand, it must be admitted that any naturally integer linear programming solutions at Step 8 have nothing to do with surrogate constraints.

The RIP30C code referred to above has subsequently undergone extensive modification. Among the new features used routinely are the following: a dual linear programming subroutine which resolves degeneracies more efficiently and permits column-generation and reduced storage requirements for large problems; penalties of the type discussed in §3.1.4 for fathoming and for selecting a separation variable; and means for "purging" the candidate list as suggested in the remark on Step 12 in §2.4. The performance of the code is perhaps best indicated by the ratio of total time required to solve an integer program to the elapsed time upon solution of the initial linear program. For an eclectic set of 8 realistic test problems in the size range of 80–131 binary variables, this ratio currently ranges from 1.05 to 3.1, with a mean of 1.67 and a median of 1.5. The computing times themselves range from 0.95 to 7.5 secs. on an IBM 360/91. All runs were made using the standard default options of the code, with no exploitation of special structure.

3.1.6. *Other Enumerative Algorithms.* Davis [21] has implemented an algorithm which, like [29], employs both logical tests and linear programming for fathoming under a primarily LIFO search strategy. In contrast to the latter algorithm, however, he relies mostly on simple penalties rather than on strongest surrogate constraints to enhance the power of the standard fathoming tests. Davis also introduced an innovation aimed at obtaining an earlier proof of termination or, failing that, an indication that certain variables may be essentially dropped for the remainder of the search because they must assume particular values. This is done by periodically solving a linear program composed of (MIP) with the integrality requirements dropped, augmented by linear constraints which specify that only those solutions need be considered which: (a) improve on the value of the current incumbent, and (b) do not duplicate any solutions already accounted for by fathoming. An implementation for the Univac 1108 has been extensively tested. To mention just the larger problems: two project selection problems with 105 binary variables and 2 constraints were solved in 5 and $6\frac{1}{2}$ seconds; a project selection problem with 50 binary variables and 5 constraints was solved in $10\frac{1}{2}$ seconds; two machine scheduling problems with 131 binary variables and 40 constraints were solved in 15 and 20 seconds; and an investment planning problem with 11 binary variables, 288 continuous variables and 99 constraints was solved in 100 seconds.

Mitra [58] has done comparative studies of the relative effectiveness of a number of variants of Dakin's basic plan. The variants consist of five alternative options for Step 3 and six alternative options for the choice of separation variable at Step 11. Presumably Step 7 uses some kind of penalties, although this is not spelled out. Computational experience is cited with 9 practical problems, many of them very large (thousands of constraints and hundreds of 0-1 variables). However, no problem with more than 40 0-1 variables was solved to optimality. This makes it difficult to draw any firm conclusions from the 17 reported runs. The reader should consult the original paper, as an adequate summary would not be possible in this limited space.

MARISABETH is a mixed integer code built by Shell Berre-Paris for the Univac 1108. The algorithm is of the branch-and-bound type, but details have not been made public. The implementation is based on the Marie-Claire linear programming code of Univac France. Performance data [43] are given in Table 2. In this table, NR stands for the number of rows, NC (NB) for the number of continuous (0-1) variables, $T1$ for the time to solve the initial linear program, $T2$ for the total computing time, and NCP for the number of candidate problems examined at Step 5. Problem 1 is concerned with oil refinery planning, problems 2–5 with multi-period investments, problems 6–8 with scheduling oil product deliveries by rail, and problems 9 and 10 with coastal shipping of oil products.

3.2. *Benders Decomposition*

Bender's approach [13] to mixed integer programming rests on an equivalent version of (MIP) in terms of its integer variables alone. The equivalent version is obtained in three stages. The first stage is to "project" (MIP) onto the space of its x-variables as follows:

(MIP-1) $\text{Minimize}_{x \geq 0, \text{ integer }} [cx + \text{Infimum}_{y \geq 0} \{dy \mid Dy \leq b - Cx\}]$.

Notice that evaluating the objective function requires solving the linear programming subproblem

(SUB) $\text{Minimize}_{y \geq 0} \, dy$ subject to $Dy \leq b - Cx$.

The second stage is to dualize (SUB), so that (MIP-1) becomes

(MIP-2) $\text{Minimize}_{x \geq 0, \text{ integer }} [cx + \text{Supremum}_{u \geq 0} \{u(Cx - b) \mid d + uD \geq 0\}]$.

TABLE 2
Computational Results with MARISABETH

Problem	NR	NC	NB	$T1$	$T2$	NCP
1	156	98	25	22 sec.	3 min. 43 sec.	195
2	671	803	48	7 min. 9 sec.	14 min. 55 sec.	121
3	146	317	18	2 min. 8 sec.	4 min. 38 sec.	61
4	280	252	54	32 sec.	17 min. 52 sec.	791
5	355	769	14	4 min. 13 sec.	18 min. 12 sec.	67
6*	885	260	584	6 min. 17 sec.	68 min.	1161
7*	1254	254	839	17 min.	43 min.	889
8*	762	174	638	11 min.	75 min.	344
9*	1581	485	282	1 min. 32 sec.	17 min.	107
10*	2011	702	294	1 min. 23 sec.	2 min. 20 sec.	39

* The solutions of these problems are suboptimal, to an unspecified degree of approximation.

In terms of the extreme points $\{u^1, \cdots, u^p\}$ and extreme rays $\{u^{p+1}, \cdots, u^{p+q}\}$ of the feasible region of the dual of (SUB), this problem can be written

(MIP-3)
$$\text{Minimize}_{x \geq 0,\ \text{integer}}\ [cx + \text{Max}\ \{u^1(Cx - b), \cdots, u^p(Cx - b)\}]$$
$$\text{subject to}\quad u^j(Cx - b) \leq 0, \qquad j = p + 1, \cdots, p + q.$$

The third and final stage is to linearize (MIP-3) by expressing the maximum in the objective function as a least upper bound:

$$\text{Minimize}_{x \geq 0,\ \text{integer};\ x_0}\ cx + x_0$$

(MIP-B)
$$\text{subject to}\quad x_0 \geq u^j\,(Cx - b), \qquad j = 1, \cdots, p,$$
$$u^j(Cx - b) \leq 0, \qquad j = p + 1, \cdots, p + q.$$

See the original paper by Benders or [8] or [30], [31] for further details. The equivalence between (MIP) and (MIP-B) which justifies solving one problem via the other is this: if (x_0^*, x^*) is optimal in (MIP-B), then x^* is also optimal in (MIP) in the sense that solving (SUB) with $x = x^*$ will yield y^* such that (x^*, y^*) is optimal in (MIP).

The advantage of (MIP-B) over (MIP) is that it reduces the number of continuous variables to one (x_0); but the disadvantage is that it has many constraints that are only implicitly known. No one would seriously propose computing $u^1, \cdots, u^{p+q}$ explicitly before attempting to solve (MIP-B). Consequently, this problem is an excellent candidate for relaxation by omitting most of the constraints and generating them only as needed. This can be done within our general framework by specializing it as follows (read (MIP-B) for (MIP) at Steps 1, 2 and 12).

Step 1. Standard.

Step 2. Standard.

Step 3. A formality (this step will be executed only once).

Step 4. Omit all of the $p + q$ constraints involving u (bound x and x_0 if necessary during the early iterations—see the remark on Step 4 in §2.4).

Step 5. Solve (CP_R) optimally.

Step 6. Omit.[4]

Step 7. Standard.

Step 8. Test the feasibility of the optimal solution $(\hat{x}, \hat{x}_0)$ just obtained for (CP_R) in (MIP-B) by solving (SUB) with $x = \hat{x}$. Denote the optimal value by $v(\hat{x})$. There are three possible outcomes:

    (i) (SUB) is infeasible. Fathoming Criterion 3 is not satisfied, but the corresponding dual solution generates a violated constraint of (MIP-B). Go to Step 9.

    (ii) (SUB) is feasible and $v(\hat{x}) > \hat{x}_0$. Fathoming Criterion 3 is not satisfied, but the corresponding dual solution generates the most violated constraint of (MIP-B). Furthermore, a feasible solution of (MIP-B) is $(\hat{x}, x_0 = v(\hat{x}))$; if the value of this solution is less than Z^*, make it the new incumbent and reset Z^*. Go to Step 9.

    (iii) (SUB) is feasible and $v(\hat{x}) \leq \hat{x}_0$. Fathoming Criterion 3 is satisfied; make $(\hat{x}, \hat{x}_0)$ the new incumbent and go to Step 12.

Step 9. Always go to Step 10.

---

[4] Assuming that (MIP) has a finite optimal solution, $F(CP_R) \neq \varnothing$.

Step 10. Tighten the relaxation by adding the violated constraint just determined.

Step 11. Omit.

Step 12. Standard.

This approach, which is obviously finitely convergent, has the advantage of providing both lower and upper bounds on the optimal objective function value. A lower bound is provided each time Step 5 is executed, and an upper bound each time outcome (ii) obtains at Step 8. The two bounds approach one another as the calculations proceed, and coincide at termination.

Recent computational experience is cited by [2], [19], [45] and [51]. Aldrich's code, MIDAS-2, actually solves a somewhat weaker approximate version of the relaxed problem at Step 5. This diminishes the computational effectiveness of the code, which is implemented within IBM's Mathematical Programming System for the 360/65. MPS is used to solve the linear program at Step 8, and the integer program at Step 5 is solved by an "enhanced" version of Balas' additive algorithm [4]. The largest problem solved was of the investment planning type and had 378 constraints, 1326 continuous variables and 24 integer variables; it was solved in 4 iterations (executions of Step 5—no time given). Experience is also cited for a warehouse location problem with 39 constraints, 53 continuous variables and 14 binary variables. It was solved in 4.46 minutes and 12 iterations.

Childress cites experience with the mixed integer programming option available with Bonner and Moore's Functional Mathematical Programming System. Again a Balas-class algorithm was used for Step 5. The following two refinery design models were the largest for which computational experience was reported. The first model had 288 constraints, 390 continuous variables, and 56 binary variables, and was solved to within $\frac{1}{3}$ of 1 % of optimum in 18.13 minutes on a Univac 1108. The second model, with 515 constraints, 851 continuous variables and 48 binary variables, took 19.17 minutes to within 1.3 % of optimum.

Inman and Manne have undertaken an implementation for problems in which all of the integer variables are binary and subject to multiple choice type constraints. They are specifically interested in project evaluation problems for economic or industrial development, where the objective is to decide which of a number of possible combinations of projects to undertake. The multiple choice structure makes it possible, for problems of moderate size, to carry out the optimization at Step 5 by complete enumeration. Step 8 was accomplished using IBM's Mathematical Programming System. Computational experience is cited for one problem that dealt with plant size, time-phasing, and process choice for the steel, oil and electricity sectors of Mexico; there were 350 continuous variables, 27 binary variables, and 275 constraints. Two versions of the problem were solved on an IBM 360/67. After an initial simplex solution, one took 10 iterations and 5.30 minutes to reach optimality, and the other 25 iterations and 10.50 minutes. Another problem dealt with the location of electricity generating plants and transmission lines in Mexico; there were 622 continuous variables, 22 binary variables, and 531 constraints. After an initial simplex solution, this took 16 iterations and 28.64 minutes. Still another facility location problem had 98 continuous variables, 22 binary variables and 40 constraints. After an initial simplex solution, it took 18 iterations and 5.22 minutes.

Some earlier computational results by B. R. Buzby for a class of nonlinear distribution problems are given in [47]. Very recent results for a large multicommodity distribution problem are given in [32], but are not quoted here because the implementation is so highly specialized.

A possibly useful variation of the classical Benders Decomposition approach arises if one also relaxes the integrality requirements at Step 4. The advantage of doing this is that Step 5 can then be done by ordinary linear programming, at the cost, however, of adding additional machinery to cope with the need for integrality. Cutting-planes and enumeration are the two main choices for this extra machinery. If enumeration is selected (as in [49] and [2]), then the following alterations are necessary in the previous outline of Benders Decomposition: Step 3 is no longer a mere formality, as it will be executed more than once; Step 4 should drop all integrality requirements on x, but should retain those of the $p + q$ constraints which have so far been explicitly generated at Step 8; Step 6 cannot be omitted; Step 8 should check $\hat{x}$ for integrality first (there is no need to solve (SUB) if $\hat{x}$ is not integer); Step 9 should persist to Step 10 if and only if a violated constraint of (MIP-B) was just generated at Step 8; and Step 11 cannot be omitted. These changes also suffice to handle the case where some but not all of the integrality requirements on x are dropped at Step 4—except, of course, that Step 5 then requires a mixed integer rather than ordinary linear programming routine.

To sum up this variant of Benders Decomposition in a few words, it could correctly be called an enumerative algorithm for (MIP-B) with the $p + q$ constraints relaxed at Step 4 and generated as needed at Step 8.

3.3. *Cutting-Plane Algorithms*

Historically, this was the first general approach taken to solving integer programs. The foundations were laid by Gomory in a series of well-known papers, and there is now an extensive literature on the subject [9]. The classical pattern for cutting-plane algorithms can be described in terms of the general framework as follows.

Step 1. Standard.

Step 2. Standard.

Step 3. A formality (this step will be executed only once).

Step 4. Relax all integrality requirements.

Step 5. Solve by a linear programming algorithm.

Step 6. Standard.

Step 7. Omit.

Step 8. Standard.

Step 9. Always go to Step 10.

Step 10. Tighten the current relaxation by adding a cutting-plane.

Step 11. Never occurs.

Step 12. Standard.

Notice that the approach is based entirely on successively improved relaxations, with no use whatever made of separation. Thus termination occurs the first time that either Fathoming Criterion 1 or 3 is satisfied. It follows that the cutting-plane approach has the disadvantage of not ordinarily providing a feasible solution to the original problem until termination.

The "cutting-planes" referred to at Step 10 are, of course, linear constraints. Thus $(CP_R{}^k)$, the linear program to be solved at the kth execution of Step 5, consists of the original integer linear program with the integrality requirements dropped and $k - 1$ new linear constraints added. One infers directly from the general outline given above that the new "cuts" must be such that

$$(3) \qquad F(CP_R{}^1) \supset F(CP_R{}^2) \supset \cdots \supset F(CP_R{}^k) \supseteq F(CP).$$

That is, each new cut must properly tighten the previous relaxation, and yet still yield a valid relaxation of the (one and only) candidate problem itself. In other words, each new cut must lop off some of the feasible region of the current linear program without also lopping off any feasible integer solutions of the original integer program. It follows from (3) that

$$(3) \qquad v(CP_R^1) \leqq v(CP_R^2) \leqq \cdots \leqq v(CP_R^k) \leqq v(CP).$$

There are many ways of deriving cuts satisfying (3). A few recent references are [7], [18], [35], [37], [77]. To be useful, however, the cuts should have certain additional desirable properties. For instance, it is customary for each new cut to cut away the solution just found at Step 5, for otherwise that solution would still be valid at the next execution of Step 5. Another desirable property is that at most a finite number of cuts will be necessary in order to find an optimal solution of the given problem or discover that none exists. Still more desirable is the property that the cuts actually are faces of the smallest convex polytope containing $F(CP)$, namely the convex hull of the feasible integer solutions of the original problem. This elusive object is the ultimate objective of considerable current research, e.g., [38].

Almost all of the computational experience with cutting-plane algorithms reported in the literature has been confined to very small problems. See [9], [72] and [57] for reviews and evaluations of these efforts. A conspicuous exception is the work of Glenn Martin at Control Data Corporation [53], [54], [55], who has had outstanding success with certain types of problems having a strong tendency toward natural integrality and basic determinants of moderate size (in the associated continuous LP problem). Problems relating to networks, the traveling salesman problem, and generalized set covering problems tend to fall in this category. By far the most numerous and successful applications in Martin's experience have been crew scheduling problems, many hundreds of which he has successfully solved during the past 8 years [56]. He reports that a typical "moderate" sized crew scheduling problem of about 100 inequality constraints and 4,000 binary variables takes on the order of 10 minutes to solve on a CDC 3600. A typical larger problem of 150 rows and 7,000 binary variables can often be solved with a few cuts in on the order of 40 minutes of CDC 3600 time.

3.4. *Group Theoretic Algorithms*

The group theoretic approach was initiated by Gomory [36] and has been carried forward primarily by Shapiro [63], [64], [65]. Other recent contributions have been [34], [68], [75], and [76]. The present discussion follows Garry and Shapiro [39].

The group theoretic approach has been applied almost exclusively to the pure integer programming problem; that is, to (MIP) without any continuous variables (y). Writing $A = [C \mid I]$ to introduce slack variables, we have

$$(4) \qquad \text{Minimize } cx$$
$$\text{subject to } Ax = b, \qquad x \geqq 0 \quad \text{and} \quad \text{integer.}$$

The first step in the group theoretic approach is to transform this into an equivalent form. Let B be any dual feasible basis for (4) regarded as a linear program, and reorder the variables if necessary so that $A = [B \mid R]$. Then (4) is equivalent to the problem

$$(4A) \qquad \text{Minimize } \bar{c}x_R$$
$$\text{subject to } x_B = \bar{b} - \bar{A}x_R \geqq 0 \quad \text{and integer}, \qquad x_R \geqq 0 \quad \text{and integer}$$

where x_B and x_R denote the basic and nonbasic variables, respectively, and

$$\bar{b} = B^{-1}b,$$

$$\bar{A} = B^{-1}R,$$

$$\bar{c} = c_R - c_B B^{-1}R \geqq 0.$$

Problem (4A) is referred to as a *correction problem* since the object is to find the cheapest nonnegative integer correction x_R which renders x_B nonnegative and integer. If B is an optimal basis, then x_B is already nonnegative when $x_R = 0$.

The major relaxation employed is to drop the $x_B \geqq 0$ conditions. The basic variables, though still required to be integer, are thus allowed to become negative. This relaxation of (4A) can then be shown to be equivalent to the following group problem:

(GP)

Minimize $\bar{c}x_R$

subject to $A^*x_R = b^*$ (mod D), $x_R \geqq 0$ and integer,

where

$$D = |\det B|,$$

$$A^* = D\{\bar{A} - [\bar{A}]\},$$

$$b^* = D\{\bar{b} - [\bar{b}]\},$$

and the square brackets denote integer parts. Thus $[\bar{A}]$ is a matrix each of whose entries is the largest integer not greater than the corresponding entry in $\bar{A}$, and similarly for $[\bar{b}]$. Problem (GP) can be interpreted as an optimization over the finite Abelian group G generated by the columns of A^*, and can be solved by a special shortest path algorithm.

The group problem (GP) can be used to fathom candidate problems in exactly the same way linear programs are used in the algorithms of §3.1. The method of creating candidate problems used in [39] is considerably different from any discussed so far, however, and so will be described briefly. For simplicity, assume that the integer variables have upper bounds greater than one. Let $\tilde{x}_R = (\tilde{x}_1, \cdots, \tilde{x}_n)$ be an arbitrary nonnegative correction and define

$$j(\tilde{x}_R) = \min \{j \mid \tilde{x}_j > 0\}.^5$$

The candidate problem associated with $\tilde{x}_R$ is

Minimize $\bar{c}(\tilde{x}_R + u)$

(CP − $\tilde{x}_R$)

subject to $x_B = \bar{b} - \bar{A}(\tilde{x}_R + u) \geqq 0$ and integer,

$u_j \geqq 0$ and integer for $j = 1, \cdots, j(\tilde{x}_R)$, $u_j = 0$ for $j = j(\tilde{x}_R) + 1, \cdots, n$.

This candidate is relaxed to a group problem by dropping the $x_B \geqq 0$ condition. The list of candidate problems is separated according to "level number." Starting at $K = 0$ the corrections at level K are those corrections x_R with $\sum_{j=1}^{n} x_j = K$. If an arbitrary correction $\tilde{x}_R$ at level K cannot be fathomed, then it is replaced by the cor-

---

[5] If $\tilde{x}_R = 0$, then define $j(\tilde{x}_R) = n$.

rections

$$\tilde{x}_R + e_j \quad \text{for} \quad j = 1, \cdots, j(\tilde{x}_R)$$

each of which is at level $K + 1$. (Here e_j denotes the jth unit vector.) This effects a separation of $(CP - \tilde{x}_R)$ into $j(\tilde{x}_R)$ new candidate problems.

The algorithm of [39] can now be related to the general framework.

Step 1. The given problem (4) is transformed into a correction problem (4A). Any dual feasible basis can be used, but the usual procedure is to use an optimal basis. Let $\tilde{x}_R = 0$.

Step 2. Standard.

Step 3. The selection rule employed is called "directed search". Each candidate problem is assigned a "plausibility value". This is a lower bound on its minimal value, computed at the time of creation from the group problem solution. The candidate selected is the one with the lowest lower bound from among a subset (those in main storage) of the pending candidates.

Step 4. Relax the candidate problem $(CP - \tilde{x}_R)$ to a group problem by dropping the nonnegativity conditions on the basic variables.

Step 5. For the relaxation performed at Step 4, a known algorithm is available to set up a canonical representation of the group and solve the problem. For the other kinds of relaxation performed at Step 10, the appropriate algorithms are assumed to be available.

Step 6. Standard.

Step 7. Standard.

Step 8. Standard.

Step 9. Heuristic rules may be used so that "a reasonable computational effort" is made to fathom the current candidate problem.

Step 10. Several alternative types of relaxation are proposed for this step. These include Lagrangian group problems [66], cutting planes (§3.3), and surrogate constraints (§3.1.5).

Step 11. The separation technique is as described earlier. The group (or other) problem solution is used to compute lower bounds on the minimal values of the new candidate problems. These are referred to as "plausibility values."

Step 12. Standard.

Step 10 could also incorporate a variety of heuristics and adaptive devices to choose a "suitable" relaxation for each candidate problem, depending on its algebraic structure.

Some computational results are given. For example, several media selection problems with about 75 constraints and 150 integer variables were solved in under a minute on a Univac 1108.

Some efforts have been made to extend the theory to mixed integer programming [75]. The computational difficulties encountered, however, seem formidable. More promising are attempts to synthesize the group approach with Benders Decomposition which separates the pure integer and continuous parts of a mixed integer problem.

IV. Summary and Prognosis

In this final section, we would like to summarize very briefly the current state-of-the-art of integer linear programming, and offer some opinions as to promising directions for the future.

Enumerative algorithms have received the lion's share of attention in recent years, especially as measured by new implementations. This is due partly to disillusionment with the erratic computational performance of the early cutting-plane algorithms, and partly to the publication of such influential papers as [46], [50], [4], and [24]. Recent computational experience seems to vindicate this emphasis. A number of existing codes are quite reliable in obtaining optimal solutions within a short time for general all-integer linear programs with a moderate number of variables—say on the order of 75—and well into the hundreds of variables for problems of more special structure. Mixed integer linear programs of practical origin with up to half a hundred integer variables and a thousand or two continuous variables and constraints are tractible for several modern implementations based on a production quality linear programming system. Problems of this size with several hundred integer variables have been undertaken, but usually the result has been a suboptimal feasible solution (often an acceptably good one). All of the successful general purpose codes use linear programming at Step 5, almost all of them use the penalty idea in several ways, and many use some sort of compromise between strict LIFO and Priority at Step 3 and permit several options for selecting a separation variable at Step 11.

The available computational experience with Benders-type algorithms, although not nearly so voluminous as that for enumerative algorithms, does suggest that this approach may be competitive for mixed integer linear programs. The implementations to date, however—at least the ones of which we are aware—do not achieve the full potential inherent in the approach because they use only a relatively rudimentary enumerative all-integer algorithm at Step 5. Thus the computing time spent in Step 5 is longer than necessary, and may account for the fact that only a few dozen integer variables have so far been handled successfully. The incorporation of a modern integer algorithm into Step 5 would be a programming task of considerable magnitude, but the result might well be a code capable of routinely handling one or two hundred integer variables and a few thousand continuous variables and constraints. Such success will almost certainly come to pass if the number of iterations (executions of Step 5) required to find a near optimal solution tends to remain moderate as problem size increases. Currently there is inadequate computational experience to resolve this crucial issue, but the available evidence suggests some optimism. Often only a surprisingly small number of iterations has been necessary, and the number of iterations has usually not increased in proportion to the number of logical alternatives to be accounted for, or even in proportion to the number of integer variables (see especially Table 5 in [51] and §7.3.5 of [47]).

One should also keep in mind where Benders Decomposition is concerned the portentous fact that the integer and continuous parts of the problem are treated in a way that preserves the structure of the continuous part. If the integer variables represent different configurations of a system, for example, then the integer optimization at Step 5 selects a new configuration to be tried and Step 8 optimizes the system with this configuration. The latter optimization can sometimes be done very efficiently by a special purpose algorithm. An instance' of this occurs in some problems of optimal multicommodity distribution system design [32], where Step 8 breaks apart into a number of independent ordinary transportation problems (one for each commodity).

As for the cutting-plane and group theory approaches, it appears that unless significant new theoretical advances are made it may be best to marry them with an enumerative approach. Since [39] explained so clearly how this can be done with the group theory approach, we confine our remarks here to the synthesis of the cutting-plane approach with enumeration, a topic that has thus far received much less attention than it deserves (cf., p. 897 of [49] and p. 243 of [9]).

The idea is very simple. Just take a Dakin-type enumerative algorithm, say, and equip Step 10 with the means of generating cutting planes for the current candidate problem. A plausible rule for Step 9 would then be: for each candidate problem, persist to Step 10 up to a certain number of times in succession, or until the rate of improvement due to adding new cuts falls below some threshold value. This rule is appealing because cutting-plane algorithms typically make the greatest rate of progress during the early iterations (e.g., [71]). This means that there is a chance of actually solving within a few cuts a candidate problem unfathomable by standard means, and in any case a better lower bound on its optimal value is generated each time a cut is added—which improves the chances of activating the second fathoming criterion. This synthesis, incidentally, can be viewed as a natural extension of the use of penalties at Step 7, for penalties are analogous to estimating the change induced in the optimal value of the relaxed candidate problem by adding a single very simple cut (cf., the conclusion of §3.1.4). Obviously, adding more than one cut will yield improved estimates of the optimal value of the candidate problem.

In addition to accommodating syntheses between traditionally distinct approaches, the general framework of Figure 1 also provides a checklist of opportunities for improving any algorithm falling within its scope. It is particularly valuable to refer to this checklist when it is desired to specialize an algorithm to a particular class of problems such as fixed charge, project selection, scheduling, etc. Can a prior analysis of the problem reveal some natural preliminary separations that ought to be made, and perhaps arranged in a certain desired order for processing? If so, this prior analysis should be built into Step 1. Is there a better way of telling how "promising" a candidate problem is other than by the usual bound that may be associated with it? If so, then Step 3 can be modified accordingly. Is there a different kind of relaxation that could be employed at Step 4 or 10 which would make the resulting relaxed candidate problems easier to solve at Step 5 without unduly compromising the power of Steps 6, 7 and 8, or that would enhance the power of these steps without being too much more difficult to solve at Step 5? Does the problem structure permit improved sufficient conditions for Steps 6 and 7? Can indicators be designed to predict with some degree of reliability when persistence at Step 9 is warranted? What class of separations should be allowed at Step 11, and by what rule should members of the class be chosen, so as to most enhance the fathoming tractability of the descendants of a candidate problem?

Finally, we hazard a guess that the field of integer programming may be profoundly affected by the advent of highly parallel fourth generation computers such as the ILLIAC IV. The reason is that the ability to process several candidate problems simultaneously gives rise to a new range of possible enumerative strategies quite different from those used to date. We hope that algorithm designers will take the initiative in exploiting these new possibilities.

References

1. AGIN, N., "Optimum Seeking with Branch and Bound," *Management Science*, Vol. 13, No. 4 (December 1966), pp. B-176-185.
2. ALDRICH, D. W., "A Decomposition Approach to the Mixed Integer Problem," Doctoral Dissertation, School of Industrial Engineering, Purdue University (1969).
3. APPELGREN, L. H., "Integer Programming Methods for a Vessel Scheduling Problem," Report No. R35 (January 1970), Institute for Optimization and Systems Theory, The Royal Institute of Technology, Stockholm.

4. BALAS, E., "An Additive Algorithm for Solving Linear Programs with Zero-One Variables," *Operations Research*, Vol. 13, No. 4 (July–August 1965), pp. 517–546.

5. ——, "Discrete Programming by the Filter Method," *Operations Research*, Vol. 15, No. 5 (September–October 1967), pp. 915–957.

6. ——, "A Note on the Branch and Bound Principle," *Operations Research*, Vol. 16, No. 2 (March–April 1968), pp. 442–445.

7. ——, "Intersection Cuts from Maximal Convex Extensions of the Ball and Octahedron," MSRP No. 214 (August 1970), Graduate School of Industrial Administration, Carnegie-Mellon University, Pittsburgh.

8. BALINSKI, M. L., "Integer Programming: Methods, Uses, Computation," *Management Science*, Vol. 12, No. 3 (November 1965), pp. 253–313.

9. —— AND SPIELBERG, K., "Methods for Integer Programming: Algebraic, Combinatorial and Enumerative," in J. S. Aronofsky (ed.), *Progress in Operations Research, Vol.* III, Wiley, New York, 1969.

10. BEALE, E. M. L., "Survey of Integer Programming," *Operational Research Quarterly*, Vol. 16, No. 2 (June 1965), pp. 219–228.

11. —— AND SMALL, R. E., "Mixed Integer Programming by a Branch and Bound Technique," in W. A. Kalenich (ed.), *Proceedings of the IFIP Congress 1965*, Vol. 2, Spartan Press, Washington, D. C., 1965.

12. —— AND TOMLIN, J. A., "Special Facilities in a General Mathematical Programming System for Non-Convex Problems Using Ordered Sets of Variables," in J. Lawrence (ed.), *Proceedings of the Fifth International Conference of Operational Research*, Venice, Tavistock Publications, London, 1969.

13. BENDERS, J. F., "Partitioning Procedures for Solving Mixed-Variables Programming Problems," *Numerische Mathematik*, Vol. 4 (1962), pp. 238–252.

14. BÉNICHOU, M., GAUTHIER, J. M., GIRODET, P., HENTGÈS, G., RIBIÈRE, G. AND VINCENT, O., "Experiments in Mixed Integer Linear Programming," presented at the Seventh International Mathematical Programming Symposium, 1970, The Hague, Holland. To appear in *Mathematical Programming*, Vol. 1, No. 1.

15. BENNETT, J. M., COOLEY, P. C. AND EDWARDS, J., "The Performance of an Integer Programming Algorithm with Test Examples," *Australian Computer Journal*, Vol. 1, No. 3 (November 1968), pp. 182–185.

16. BERTIER, P. AND ROY, B., "Procédure de Résolution pour une Classe de Problèmes pouvant avoir un Charactère Combinatoire," *Cahiers du Centre d'Etudes de Recherche Operationnelle*, Vol. 6 (1964), pp. 202–208.

17. BLONDEAU, J. P., Private communication, May 3, 1971, Control Data Corporation, Cybernet Service Division, Houston.

18. BOWMAN, J. AND NEMHAUSER, G., "Deep Cuts in Integer Programming," Technical Report No. 8 (February 1968), Dept. of Statistics, Oregon State University, Corvallis, Oregon.

19. CHILDRESS, J. P., "Five Petrochemical Industry Applications of Mixed Integer Programming," Bonner & Moore Associates, Inc., Houston, March 1969.

20. DAKIN, R. J., "A Tree Search Algorithm for Mixed Integer Programming Problems," *Computer Journal*, Vol. 8, No. 3 (1965), pp. 250–255.

21. DAVIS, R. E., "A Simplex-Search Algorithm for Solving Zero-One Mixed Integer Programs," Technical Report No. 5 (October 1969), Dept. of Operations Research, Stanford University.

22. ——, KENDRICK, D. A. AND WEITZMAN, M., "A Branch and Bound Algorithm for Zero-One Mixed Integer Programming Problems," Development Economic Report No. 69 (1967), Center for International Affairs, Harvard University.

23. DAVIS, P. S. AND RAY, T. L., "A Branch-Bound Algorithm for the Capacitated Facilities Location Problem," *Naval Research Logistics Quarterly*, Vol. 16, No. 3 (September 1969), pp. 331–344.

24. DRIEBEEK, N. J., "An Algorithm for the Solution of Mixed Integer Programming Problems," *Management Science*, Vol. 12, No. 7 (March 1966), pp. 576–587.

25. EDNEY, M. R., Private communication, December 10, 1970, University Computing Company, London.

26. FLEISCHMANN, B., "Computational Experience with the Algorithm of Balas," *Operations Research*, Vol. 15, No. 1 (January–February 1967), pp. 153–155.

27. GATELY, D., Private communication, February 19, 1971. The model is described in "Investment Planning for the Electric Power Industry: An Integer Programming Approach,"

Research Report 7035, November 1970, Dept. of Economics, The University of Western Ontario, London, Canada.

28. GEOFFRION, A. M., "Integer Programming by Implicit Enumeration and Balas' Method," *SIAM Review*, Vol. 9, No. 2 (April 1967), pp. 178–190.

29. ——, "An Improved Implicit Enumeration Approach for Integer Programming," *Operations Research*, Vol. 17, No. 3 (May–June 1969), pp. 437–454.

30. ——, "Elements of Large-Scale Mathematical Programming," *Management Science*, Vol. 16, No. 11 (July 1970), pp. 652–691.

31. ——, "Generalized Benders Decomposition," Working Paper No. 159 (April 1970), Western Management Science Institute, UCLA. To appear in *Journal of Optimization Theory and Application*.

32. —— AND GRAVES, G. W., "Multicommodity Distribution System Design by Benders Decomposition," Western Management Science Institute, UCLA, (forthcoming 1971).

33. GLOVER, F., "A Multiphase-Dual Algorithm for the Zero-One Integer Programming Problem," *Operations Research*, Vol. 13, No. 6 (November–December 1965), pp. 879–919.

34. ——, "Integer Programming over a Finite Additive Group," *SIAM J. Control*, Vol. 7, No. 2 (May 1969), pp. 213–231.

35. ——, "Convexity Cuts," Working Paper (December 1969), School of Business, University of Texas.

36. GOMORY, R. E., "On the Relation between Integer and Non-Integer Solutions to Linear Programs," *Proc. Nat. Acad. Sci.*, Vol. 53 (1965), pp. 260–265.

37. ——, "Some Polyhedra Related to Combinatorial Problems," *Journal of Linear Algebra and Applications*, Vol. 2, No. 4 (October 1969), pp. 451–558.

38. —— AND JOHNSON, E. L., "Some Continuous Functions Related to Corner Polyhedra," RC-3311 (February 1971), IBM, Yorktown Heights, N. Y.

39. GORRY, G. A. AND SHAPIRO, J. F., "An Adaptive Group Theoretic Algorithm for Integer Programming Problems," *Management Science*, Vol. 17, No. 5 (January 1971), pp. 285–306.

40. ——, —— AND WOLSEY, L. A., "Relaxation Methods for Pure and Mixed Integer Programming Problems," WP 456–70 (April 1970), Sloan School of Management, M.I.T. To appear in *Management Science* (January 1972).

41. HAMMER, P. L. AND RUDEANU, S., *Boolean Methods in Operations Research and Related Areas*, Springer-Verlag, Berlin, 1968.

42. HELD, M. AND KARP, R. M., "The Traveling Salesman Problem and Minimum Spanning Trees," *Operations Research*, Vol. 18, No. 6 (November–December 1970), pp. 1138–1162.

43. HERVÉ, P., Private communication, October 2, 1970, Compagnie de Raffinage Shell Berre, Paris.

44. IBARAKI, T., LIU, T. K., BAUGH, C. R. AND MUROGA, S., "An Implicit Enumeration Program for Zero-One Integer Programming," Report No. 305 (January 1969), Dept. of Computer Science, University of Illinois, Urbana.

45. INMAN, R., "User's Guide to IPE," Memorandum 71-2 (February 1971), Development Research Center, International Bank for Reconstruction and Development, Washington, D. C.

46. LAND, A. H. AND DOIG, A. G., "An Automatic Method of Solving Discrete Programming Problems," *Econometrica*, Vol. 28 (1960), pp. 497–520.

47. LASDON, L. S., *Optimization Theory for Large Systems*, Macmillan, New York, 1970.

48. LAWLER, E. L. AND WOOD, D. E., "Branch and Bound Methods: A Survey," *Operations Research*, Vol. 14, No. 4 (July–August 1966), pp. 699–719.

49. LEMKE, C. E. AND SPIELBERG, K., "Direct Search Algorithms for Zero-One and Mixed-Integer Programming," *Operations Research*, Vol. 15, No. 5 (September–October 1967), pp. 892–914.

50. LITTLE, J. D. C., MURTY, K. G., SWEENEY, D. W. AND KAREL, C., "An Algorithm for the Travelling Salesman Problem," *Operations Research*, Vol. 11, No. 6 (November–December 1963), pp. 972–989.

51. MANNE, A. S., "A Mixed Integer Algorithm for Project Evaluation," Memorandum 71-3 (February 1971), Development Research Center, International Bank for Reconstruction and Development, Washington, D. C.

52. MARSTEN, R. E., "An Algorithm for the Set Partitioning Problem with Side Constraints," forthcoming Working Paper ✳181 (October 1971), Western Management Science Institute, UCLA.

53. MARTIN, G. T., "An Accelerated Euclidean Algorithm for Integer Linear Programming," in

R. L. Graves and P. Wolfe (eds.), *Recent Advances in Mathematical Programming*, McGraw-Hill, New York, 1963.

54. ——, "Solving the Travelling Salesman Problem by Integer Linear Programming," Control Data Corporation, New York, May 1966.

55. ——, "Integer Programming: Gomory Plus Ten," 38th National ORSA Meeting, Miami, November 1969.

56. ——, Private communication, May 4, 1971, Control Data Corporation, Midtown Data Center, New York.

57. MEARS, W. J. AND DAWKINS, G. S., "Comparison of Integer Programming Algorithms," The Pace Company, P.O. Box 26637, Houston. Presented at the 1968 Joint National Meeting of the Operations Research Society of America and The Institute of Management Sciences, San Francisco, May 1968.

58. MITRA, G., "Designing Branch and Bound Algorithms for Mathematical Programming," SIA Limited, 23 Lower Belgrave St., London. Presented at the Seventh International Symposium on Mathematical Programming, The Hague, Holland, September 1970.

59. MITTEN, L. G., "Branch-and-Bound Methods: General Formulation and Properties," *Operations Research*, Vol. 18, No. 1 (January–February 1970), pp. 24–34.

60. PIERCE, J. F., "Application of Combinatorial Programming to a Class of All-Zero-One Integer Programming Problems," *Management Science*, Vol. 15, No. 3 (November 1968), pp. 191–209.

61. ROY, B., BENAYOUN, R. AND TERGNY, J., "From S.E.P. Procedure to the Mixed Ophelie Program," in J. Abadie (ed.), *Integer and Nonlinear Programming*, North-Holland, Amsterdam, 1970.

62. SCHRAGE, L. AND WOILER, S., "A General Structure for Implicit Enumeration," Dept. of Industrial Engineering, Stanford University (1967).

63. SHAPIRO, J. F., "Dynamic Programming Algorithms for the Integer Programming Problem—I: The Integer Programming Problem Viewed as a Knapsack Type Problem," *Operations Research*, Vol. 16, No. 1 (January–February 1968), pp. 103–121.

64. ——, "Group Theoretic Algorithms for the Integer Programming Problem—II: Extension to a General Algorithm," *Operations Research*, Vol. 16, No. 5 (September–October 1968), pp. 928–947.

65. ——, "Turnpike Theorems for Integer Programming Problems," *Operations Research*, Vol. 18, No. 3 (May–June 1970), pp. 432–440.

66. ——, "Generalized Lagrange Multipliers in Integer Programming," *Operations Research*, Vol. 19, No. 1 (January–February 1971), pp. 68–76.

67. SOMMER, D., Private communication, March 30, 1971, Control Data Corporation, Cybernet Service Division, Minneapolis.

68. THIRIEZ, H., "Airline Crew Scheduling: A Group Theoretic Approach," Report R-69 (October 1969), Flight Transportation Laboratory, M.I.T.

69. TOMLIN, J. A., "An Improved Branch and Bound Method for Integer Programming," *Operations Research*, Vol. 19, No. 4 (July–August 1971), pp. 1070–1075.

70. ——, "Branch and Bound Methods for Integer and Non-Convex Programming," in J. Abadie (ed.), *Integer and Nonlinear Programming*, North-Holland, Amsterdam, 1970.

71. TRAUTH, C. A. AND WOOLSEY, R. E., "MESA, A Heuristic Integer Linear Programming Technique," Research Report SC-RR-68-299 (July 1968), Sandia Laboratories, Albuquerque, New Mexico.

72. —— AND ——, "Integer Linear Programming: A Study in Computational Efficiency," *Management Science*, Vol. 15, No. 9 (May 1969), pp. 481–493.

73. WAGNER, W. H., "An Application of Integer Programming to Legislative Redistricting," Crond, Inc., Wilmington, Delaware. Presented at the 34th National Meeting of the Operations Research Society of America, Philadelphia, November 1968.

74. WOILER, S., "Implicit Enumeration Algorithms for Discrete Optimization Problems," Technical Report No. 4 (May 1967), Dept. of Industrial Engineering, Stanford University.

75. WOLSEY, L. A., "Mixed Integer Programming: Discretization and the Group Theoretic Approach," Technical Report No. 42 (June 1969), Operations Research Center, M.I.T.

76. ——, "Extensions of the Group Theoretic Approach in Integer Programming," Working Paper (October 1970), Manchester Business School, University of Manchester, England.

77. YOUNG, R. D., "New Cuts for a Special Class of 0-1 Integer Programs," Research Report in Applied Mathematics and Systems Theory (November 1968), Rice University, Houston.

6. OPTIMAL SET COVERING: A SURVEY[+]

by

ROBERT GARFINKEL[++]

and

G. L. NEMHAUSER[+++]

[+]This paper is based on a chapter of the authors' forthcoming book Integer Programming, John Wiley & Sons Inc., (spring, 1972) and, in part, on the unpublished paper [1].

[++]Graduate School of Management, University of Rochester.

[+++]Department of Operations Research, Cornell University.

1.   INTRODUCTION

Suppose there is a set $I = \{1,\ldots,m\}$,   and a set $P = \{P_1,\ldots,P_n\}$ where $P_j \subseteq I$, $j \in J = \{1,\ldots,n\}$.   A   subset $J^* \subseteq J$ defines a <u>cover</u> of I if

(1)
$$\bigcup_{j \in J^*} P_j = I.$$

If, in addition,

(2) $j,k \in J^*$, $j \neq k \Rightarrow P_j \cap P_k = \phi$ (the empty set)

J^* defines a <u>partition</u> of I.   For convenience, we sometimes refer to the index set J^* as the cover or partition.

Let a cost $c_j > 0$ be associated with every $j \in J$. The total cost of the cover J^* is $\sum_{j \in J^*} c_j$.   The problem of finding a cover which has minimum cost can be written as an integer linear programming problem (ILP).

(3) $\min x_o = \sum_{j=1}^{n} c_j x_j$

(4) $\sum_{j=1}^{n} a_{ij} x_j \geq 1$ $i = 1,\ldots,m$

(5) $x_j = 0,1$ $j = 1,\ldots,n$

where

$$x_j = \begin{cases} 1 & \text{if } j \text{ is in the cover} \\ 0 & \text{otherwise} \end{cases}$$

and

$$a_{ij} = \begin{cases} 1 & \text{if} \quad i \in P_j \\ \\ 0 & \text{otherwise} . \end{cases}$$

The problem given by (3) - (5) will be called a <u>set covering problem</u> (CP).

Similarly, the <u>set partitioning problem</u> (PP) is obtained by replacing (4) with

(4a) $$\sum_{j=1}^{n} a_{ij} x_j = 1 \qquad i = 1,\ldots,m.$$

The CP is the more general of the two problems because, by changing the cost vector, any PP having a feasible solution can be transformed into a CP. In particular consider any PP defined by $A = \{a_{ij}\}$ and cost vector c. Let $t_j = \sum_{i=1}^{m} a_{ij}$ and choose any $L > \sum_{j=1}^{n} c_j$. Then

<u>Theorem 1</u> [2]   If the PP defined by A and c has a feasible solution, the CP defined by A and costs $c'_j = c_j + Lt_j$, $j = 1,\ldots,n$ has the same set of optimal solutions as the PP.

The proof of Theorem 1 is based on the fact that when L is sufficiently large, the cost of any partition is smaller than the cost of any cover which is not a partition.

Covering the Vertices of a Graph by Edges

Let $G = (V,E)$ be an undirected graph with vertex set $V = \{1,\ldots,m\}$ and edge set $E = \{e_1,\ldots,e_n\}$. The edge e_j is assigned a cost c_j, $j = 1,\ldots,n$ and the cost assigned to a subset of edges is the sum of the costs over all edges in the subset.

Two closely related problems defined on G are
a) <u>graph covering problem (GP)</u>: find a subset of edges E*

of minimum cost from among all subsets having the property
that every vertex of V is incident to at least one edge
of the subset;

b) matching problem (MP): find a subset of edges E^o of maximum cost from among all subsets having the property that
every vertex of V is incident to no more than one edge of
the subset.

The GP is clearly a CP having V = I and E = P. In fact
it is a very highly structured CP, because the matrix $A = \{a_{ij}\}$
of the GP is the vertex-edge incidence matrix of G. This
structure has yielded a powerful theory and highly efficient
algorithms for GP's and MP's. Unfortunately, these results
have not yet been extended to general covering problems, where
almost all of the applications arise.

The bulk of this paper contains a survey of applications,
theory and algorithms for the general covering problem. In the
last section, a summary of some basic results for graph covering
problems is given. References [37] - [47] give results which
fall outside of the mainstream of our presentation and therefore are not cited in the text.

2. SOME APPLICATIONS

Information Retrieval [3]

Consider the problem of retrieving information from n
files where the jth file is of length c_j, j = 1,...,n. Suppose m requests for information are received. Each unit of
information i is stored in at least one file j, indicated
by a_{ij} = 1. An optimal solution to the CP yields a subset
of files which minimizes the maximum total length which needs
to be searched in order to guarantee retrieval of all of the
information.

Disconnecting Paths in a Graph [4]

Consider a graph G = (V,E). Let I correspond to a set

of paths of the graph, and let $J = \{1,\ldots,|E|\}$.[+] If $a_{ij} = 1$ means that edge j is in path i, and c_j is the cost of removing edge j from the graph, then an optimal CP solution yields a minimum cost set of edges that disconnects all paths of I. Applications in transportation, fluid flow and electrical systems are obtained by taking the graph to represent an appropriate physical network.

Truck Deliveries [5]

Consider the problem of making deliveries to m locations by truck. There are n feasible routes to choose from and $a_{ij} = 1$ if location i is on route j. A cost c_j (possibly its length) is assigned to route j. Then an optimal solution to the PP gives a minimal cost routing which makes each delivery exactly once.

Political Districting [6], [7]

Here I corresponds to a set of basic population units (counties, census tracts, etc.). A subset P_j of I is in P if the population units form a district which meets requirements on population, contiguity, compactness, etc. In addition, it is required to have exactly K districts. If c_j is some measure of the unacceptability of district j, then an optimal solution to the PP, together with the additional constraint that $\sum_j x_j = K$, yields an optimal districting plan.

Airline Crew Scheduling [8], [9]

An airline has a set of m flight "legs", each of which requires a crew. Let $P_j \in P$ if it corresponds to a set that can be handled by a single crew. Let c_j be some measure of the cost of P_j. Depending on whether or not crew members are allowed to be passengers on certain flights, optimal solutions

---

[+]$|S|$ denotes the cardinality of the set S.

to the CP or the PP yield optimal schedules.

Coloring Problems [10]

Consider the problem of coloring a map so that no two adjacent areas have the same color. Let I represent the set of areas. A subset P_j is in P if no two elements of P_j correspond to areas having a common boundary. If all costs are unity, an optimal solution to the PP indicates the minimum number of colors needed.

There are many other applications of the CP and PP models including problems such as the optimal design of switching circuits and assembly line balancing (see references [11] - [16]).

3. REDUCTIONS

In set covering and partitioning problems, it is often possible a priori to eliminate certain rows and columns from A. Some of these reductions are valid for the CP, some for the PP, and some for both. The notation r_i is used for row i, and a_j for column j of $A = \{a_{ij}\}$.

Reduction 1 (CP,PP) - If r_i is a null vector for some i, there is no feasible solution since the ith constraint cannot be satisfied.

Reduction 2 (CP,PP) - If $r_i = e_k$ (a unit vector with one in the kth position) for some i, k, then $x_k = 1$ in every feasible solution, and a_k may be deleted. Also, every row corresponding to $t \in P_k$ may be deleted.

Since r_i cannot be covered without a_k, and every element of P_k is covered by a_k, this reduction is obvious.

Reduction 2a (PP) - In addition to the deletions of reduction 2, every column $r \neq k$ such that $a_{tr} = a_{tk} = 1$ for some $t \neq i$ must be deleted.

Since $x_k = 1$, setting $x_r = 1$ for any such column would result in $\sum_{j=1}^{n} a_{tj} x_j \geq 2$ and row t would be "overcovered."

Reduction 3 (CP,PP) - If $r_t \geq r_p$ (in a vector sense) for some rows t and p, then r_t may be deleted.

This reduction follows because every solution satisfying the pth constraint automatically satisfies the tth constraint.

Reduction 3a (PP) - In addition to deleting r_t in Reduction 3, every column k such that $a_{tk} = 1$ and $a_{pk} = 0$ must be deleted.

This follows because some column r with $a_{pr} = a_{tr} = 1$ must have $x_r = 1$ in order to cover row p. Thus, if $x_k = 1$, row t would be "overcovered."

Reduction 4 (PP,CP) - If for some set of columns S and some column k, $\sum_{j\epsilon S} a_j = a_k$ and $\sum_{j\epsilon S} c_j \leq c_k$, column k may be deleted.

Reduction 4a (CP) - If for some set of columns S and some column k, $\sum_{j\epsilon S} a_j > a_k$ and $\sum_{j\epsilon S} c_j \leq c_k$, column k may be deleted.

4. HANDLING BINARY DATA

In the set covering problem all of the data except the costs are 0,1. Since virtually all scientific computers store their information in binary, one piece of data can be stored as one "bit." For instance, the IBM 7094 has a 36 bit word, so any column or row of A with 36 bits or less can be stored as one word. In general, if the word size of the computer is k bits and a column (row) has q entries, it takes $<q/k>$ words to store the column (row).[+] Thus much larger matrices can be saved than would be the case in a general ILP.

Another handy feature of binary storage is that certain calculations can be done using the logical AND and logical OR statements of computer assembly languages. These two operations are performed on vectors as follows:

[+] $<a>$ denotes the smallest integer equal to or greater than the real number a.

AND

$$a \circ b = c$$

$$c_j = \begin{cases} 1 & \text{if } a_j = b_j = 1 \\ 0 & \text{otherwise} \end{cases}$$

OR

$$a <> b = c$$

$$c_j = \begin{cases} 0 & \text{if } a_j = b_j = 0 \\ 1 & \text{otherwise} \end{cases}$$

The reductions of the last section are handled quite easily by these calculations. For example, in reduction 3, $r_t \geq r_p$ is equivalent to $r_t \circ r_p = r_p$.

Also, some of the algorithms described later perform the operations $\cup$ and $\cap$ on sets where the sets are represented by binary vectors. Clearly these set operations correspond to the operations $<>$ and $\circ$ respectively on binary vectors.

5. SOME EXTREME POINT PROPERTIES

In this section relationships between covering and partitioning problems and some related linear programs (LP's) are examined. The LP obtained by replacing (5) with

(5a) $x_j \geq 0 \qquad j = 1, \ldots, n$

in the CP is denoted by CP', and in the PP by PP'.

Redundant Solutions to the CP

Let x' be any feasible solution to the CP, and $J' = \{j \mid x'_j = 1\}$ be the corresponding cover. A column $j* \epsilon J'$ is said to be redundant, if $J' - \{j*\}$ is also a cover. If a cover contains one or more redundant columns, it is also termed redundant. A cover which is not redundant is called prime. Column j* is redundant with respect to the cover given by J' if and only if

(6) $\sum_{j \epsilon J'} a_{ij} \geq 2 \quad \text{for all } i \epsilon P_{j*}$.

Equivalently, $\hat{j} \epsilon J'$ is not redundant if and only if

(7) $I(\hat{j}) = \{i| \sum_{j \in J'} a_{ij} = 1, \quad i \varepsilon P_{\hat{j}}\} \neq \phi.$

Since $c_j > 0$ for all j, every optimal cover is prime.

Now let x^* be any <u>optimal</u> solution to CP'. Clearly $0 \leq x_j^* \leq 1$, $j = 1,\ldots,n$. This follows, since $c_j > 0$ and for any feasible solution x' to CP', $x_j'' = \min\{1,x_j'\}$, $j = 1,\ldots,n$ is also feasible to CP'. Given x^*, a feasible solution x^{**} to the CP can be constructed by letting

$$x_j^{**} = <x_j^*>, \quad j = 1,\ldots,n .$$

Of course, x^{**} is not necessarily optimal to the CP. In fact, the corresponding cover J^{**} may be redundant. To obtain a prime cover from J^{**} (or from any other redundant cover), drop redundant columns from J^{**} one at a time until a prime cover is found.[+] The resulting cover will be a function of the set of redundant columns dropped which is not necessarily unique.

Associating a Basis with a Prime Cover [17]

<u>Theorem 2</u> [18]   Let J' be any prime cover and x' the corresponding solution. Then x' is an extreme point of the set of feasible solutions to the CP'.

From (7) it follows that $|J'| \leq m$ and it is quite simple to establish that the columns specified by J' are linearly independent. A basis matrix B can be associated with J', by adding $m - |J'|$ columns corresponding to surplus variables s_i of the form

$$\sum_{j=1}^{n} a_{ij}x_j - s_i = 1 .$$

[+] See Section 7 for a method of obtaining prime covers with a low objective value.

The rules for selecting the basic surplus variables are:

i) Select s_i if $\sum_{j \in J'} a_{ij} \geq 2$.

ii) For every $\hat{j} \in J'$ such that $|I(\hat{j})|$ (defined in (7)) is greater than one, arbitrarily select surplus variables corresponding to any $|I(\hat{j})|-1$ elements of $I(\hat{j})$.

These rules provide exactly $m - |J'|$ surplus variables since $m =$ number of overcovered rows $+ \sum_{\hat{j} \in J'} |I(\hat{j})|$.

The matrix B obtained by the construction just given is non-singular and has the following curious property.

Theorem 3 [17] A matrix $\hat{B}$ can be derived from B by a suitable permutation of rows such that $\hat{B}^{-1} = \hat{B}$.

The appropriate permutation is obtained by letting row i of $\hat{B}$, $i = 1,\ldots,|J'|$ be the row of B which is the unit vector e_i and for $i = |J'|+1,\ldots,m$, the row of B having a -1 in column i. This yields the self-invertible or involutory matrix

$$\hat{B} = \left(\begin{array}{c|c} I_1 & 0 \\ \hline T & -I_2 \end{array} \right)$$

where I_1 is of order $|J'|$ and I_2 is of order $m - |J'|$.

Extreme Points of PP'

For a PP, Theorem 2 can be strengthened. In particular, any partition corresponds to a prime cover. Therefore

Theorem 4 [18] - If J' is a partition, then x' is an extreme point of the set of feasible solutions to PP'.

A basis B for the PP' will be termed integer if $B^{-1}b$ is a binary vector. Two bases are adjacent if they differ in exactly one column.

Theorem 5 [19] - For every feasible integer basis to PP' there are at least as many adjacent feasible integer bases as there are non-basic columns.

It should be pointed out, however, that since integer solutions will invariably be degenerate, chances are that a large number of these adjacent bases correspond to the same extreme point.

Theorem 6 [19] - Let B_1 be a feasible integer basis to PP', and B_2 be another at least as good (in objective function value) as B_1. Let J_1 and J_2 be the corresponding partitions, and Q_1 and Q_2 be the sets of non-basic columns respectively. Then B_2 can be obtained from B_1 by a sequence of at most p pivots, where $p = |Q_1 \cap J_2|$, such that each of the p solutions is feasible, integer and not worse than its predecessor.

One would hope that this theorem would lead directly to a simple algorithm. Unfortunately some of the pivots defined in the sequence may be on negative elements, and there is no obvious way to protect against cycling.

6. A CUTTING PLANE ALGORITHM FOR THE CP [17][+]

Let x' be an integer basic feasible solution to the LP represented by

$$\min x_o = y_{oo} - \sum_{j \in R} y_{oj} x_j$$

$$x_{B_i} = y_{io} - \sum_{j \in R} y_{ij} x_j \qquad i = 1, \ldots, m$$

obtained by setting $x_j = 0$, $j \in R$, where R is the index set of non-basic variables. If x* is any other feasible integer solution to the LP, then

$$x_o^* - x_o' = - \sum_{j \in R} y_{oj} x_j^* .$$

[+]The algorithm of [17] is an extension of the results in [20].

It follows that a necessary condition for $x_o^* < x_o'$, or equivalently $x_o^* \le x_o' - 1$, is

(8)
$$\sum_{j \in Q} x_j^* \ge 1$$

where $Q = \{j \,|\, y_{oj} > 0\}$.

Furthermore, $Q = \phi$ implies x' is optimal to the LP and consequently to the corresponding ILP as well.

In order for the cut (8) to be useful in solving the CP, the set Q and consequently B^{-1} must be known, since $y_{oj} = c_B B^{-1} a_j - c_j$, $j \in R$, where c_B is the cost vector for the basic variables. However, in Section 5 it was shown that if x' is a prime solution to the CP, by a suitable permutation of rows it is possible to find a corresponding basis $\hat{B}$ such that $\hat{B} = \hat{B}^{-1}$. Thus Q can be determined without any matrix inversion.

Now in Section 5, $\hat{B}$ was shown to be of the form

$$\hat{B} = \left[\begin{array}{c|c} I_1 & 0 \\ \hline T & -I_2 \end{array} \right] = \left[\begin{array}{c|c} \hat{B}_1 & \hat{B}_2 \end{array} \right]$$

where $\hat{B}_1$ is associated with original variables and $\hat{B}_2$ with surplus variables. Also,

(9)
$$c_B \hat{B}^{-1} = (c_{B_1}, 0) \left[\begin{array}{c|c} I_1 & 0 \\ \hline T & -I_2 \end{array} \right] = c_B .$$

Now, let $A = [a_1, \ldots, a_{n+m}]$ where a_{n+i} is associated with s_i, $i = 1, \ldots, m$. Also let $\hat{A}$ be derived from A by taking the same row permutation as was used in deriving $\hat{B}$ from B. It follows that

(10)
$$B^{-1} a_j = \hat{B}^{-1} \hat{a}_j \qquad j = 1, \ldots, n+m$$

since permuting rows in a matrix has the effect of permuting the corresponding columns of the matrix inverse. Thus from (9) and (10)

$$(11) \qquad\qquad c_B B^{-1} a_j = c_B \hat{B}^{-1} \hat{a}_j = c_B \hat{a}_j \ .$$

It also follows from (11), and $c_B \geq 0$, $a_{ij} \geq 0$, $j = 1,\ldots,n$ and $a_{ij} \leq 0$, $j = n+1,\ldots,n+m$ that

$$c_B \hat{a}_j \begin{cases} \geq 0 & j = 1,\ldots,n \\ \\ \leq 0 & j = n+1,\ldots,n+m \end{cases}$$

Since $y_{oj} = c_B \hat{a}_j - c_j$ and $c_j = 0$ for $j = n+1,\ldots,n+m$, Q can only contain indices corresponding to original variables. Thus (8) is of the same form as the constraints (4), and a CP with (8) added is still a CP.

Algorithm

Step 1 - Let the original CP be denoted by CP_1, let $\bar{z} = \sum\limits_{j=1}^{n} c_j$, and let $k = 1$, where k is an iteration counter. Go to Step 2.

Step 2 - Let x^k be any prime feasible solution[+] to CP_k, and let $\hat{B}$ be a corresponding involutory basis. If $cx^k < \bar{z}$, let $\bar{z} = cx^k$ and record x^k. Go to Step 3.

Step 3 - Let $Q = \{j \,|\, c_B \hat{a}_j - c_j > 0\}$. If $Q = \phi$, terminate. The last recorded solution is optimal. If $Q \neq \phi$, go to Step 4.

Step 4 - Add the constraint $\sum\limits_{j \in Q} x_j \geq 1$ to CP_k. Let $k = k+1$ and go to Step 2.

Finiteness of the algorithm is easily shown, since at least one feasible solution to the CP is eliminated at every iteration. An upper bound on the number of such solutions, and consequently

---

[+] See Section 7 for a method of obtaining prime solutions with a low objective value.

on the number of iterations is 2^n.

Step 2 of the algorithm is vague. However, there are a number of simple ways of finding feasible (although possibly redundant) CP solutions. These can then be reduced to prime solutions as explained in Section 5. Some methods are:

<u>Method 1</u> - Set $x_j = 1$, $j = 1,\ldots,n$. Of course, this solution is feasible to any CP_k.

<u>Method 2</u> - If x is feasible to CP_k then $x + e_t$ is feasible to CP_{k+1} for any $t \in Q$, where e_t is the tth unit vector.

<u>Method 3</u> - If x is an optimal solution to CP'_k, then $x' = (<x_1>,\ldots,<x_n>)$ is feasible to CP_k.

Method 3 has considerable appeal. The optimal solution of the LP serves not only to provide a feasible integer solution, but provides a lower bound on the value of the optimal CP solution. Of course, if an optimal LP solution is integer, the CP is solved.

Computational Results

This algorithm using Method 3 has been programmed for the IBM 7094. To illustrate the running times, problems with random costs, density (number of ones/m·n) = .21, m = 30 and n = 90 took on the average 7.64 seconds.

7. AN ENUMERATION ALGORITHM FOR THE CP [2]

Impressive computational results for CP's have been obtained in [21] using a general 0-1 implicit enumeration algorithm with surrogate constraints.[+] Problems with m = 30 and n = 90 were solved on the average in 0.17 minutes on an IBM 7044.

In [2] an implicit enumeration algorithm has been designed especially for covering problems. The procedure given in [2] is described below assuming that the reader is familiar with the basic

---

[+]See also references [22-24]

ideas of implicit enumeration.

A partial solution to the CP given by (3), (4) and (5) is obtained by an arbitrary assignment of binary values to a subset of the variables. Specifically, the kth partial solution is given by

$$S_k^+ = \{j \mid x_j = 1\}$$

$$S_k^- = \{j \mid x_j = 0\}$$

$$S_k = S_k^- \cup S_k^+$$

$$F_k = J - S_k .$$

Given the kth partial solution, the reduced problem in the remaining variables x_j, $j \in F_k$, is

$$\min z_k = \sum_{j \in F_k} c_j x_j + \sum_{j \in S_k^+} c_j$$

(12)
$$\sum_{j \in F_k} a_{ij} x_j \geq 1 - \sum_{j \in S_k^+} a_{ij}, \quad i = 1, \ldots, m$$

$$x_j = 0, 1 \quad j \in F_k .$$

Letting

$$s_i = 1 - \sum_{j \in S_k^+} a_{ij}, \quad i = 1, \ldots, m$$

$$Q_k = \{i \mid s_i = 1\}$$

and $\underline{z}_k = \sum_{j \in S_k^+} c_j$, (12) can be written as the CP

$$\min z_k = \sum_{j \in F_k} c_j x_j + \underline{z}_k$$

(13)
$$\sum_{j \in F_k} a_{ij} x_j \geq 1 \quad i \in Q_k$$

$$x_j = 0, 1 \quad j \in F_k$$

Let $\bar{z}$ be the value of the best known feasible solution to the original CP. Now consider the CP' associated with (13) and the nature of its solution. There are three possibilities:

    i) The CP' has an optimal, non-integer solution given by (z_k^*, x^*);

    ii) The CP' has an optimal, integer solution given by (z_k^*, x^*). Note that in the trivial case $Q_k = \phi$, the appropriate solution is $(\underline{z}_k, 0)$.

    iii) The CP' has no feasible solution. This can only occur if $\sum\limits_{j \epsilon F_k} a_{ij} = 0$ for some $i \epsilon Q_k$. In this case, we let $z_k^* = \infty$.

    In cases ii and iii backtrack and replace $\bar{z}$ by $\min(\bar{z}, z_k^*)$. In case i, if $<z_k^*> \geq \bar{z}$, backtrack. If $<z_k^*> < \bar{z}$, the kth partial solution is not fathomed. However, a feasible solution x' to the original CP is obtained by letting $x_j' = <x_j^*>$, $j \epsilon F_k$, $x_j' = 1$, $j \epsilon S_k^+$, $x_j' = 0$, $j \epsilon S_k^-$. From x' a prime solution x^k can be obtained. Since the algorithm is enumerative, it is important to achieve tight upper bounds at every partial solution. To this end a heuristic procedure is used which attempts to produce a prime solution having a low objective value. It proceeds as follows:

Heuristic for Producing a Prime Solution

Step 1. Let $J' = \{j \,|\, x_j' = 1, j \epsilon F_k\}$. Let $J^k = \{j \,|\, x_j^k = 1, j \epsilon F_k\} = \phi$. Let $I^k = \{i \,|\, \sum\limits_{j \epsilon J^k} a_{ij} = 0, i \epsilon Q_k\} = Q_k$. Go to Step 2.

Step 2. Define $f(j) = c_j / \sum\limits_{i \epsilon I^k} a_{ij}$, $j \epsilon J' - J^k$ and choose j^* such that $f(j^*) = \min\limits_{j \epsilon J' - J^k} f(j)$. Let $J^k = J^k \cup \{j^*\}$. If $I^k \neq \phi$, repeat Step 2. If $I^k = \phi$, terminate. The solution x^k to (13) is given by $x_j^k = 1$, $j \epsilon J_k$, $x_j^k = 0$ otherwise.

    Intuitively, the heuristic attempts to include those variables which have the minimum cost per constraint satisfied.

Once x^k is obtained, replace $\bar{z}$ by $\min(\bar{z}, z^k)$, where $z^k = \sum_{j \in J^k} c_j + \underline{z}_k$. Branch to the partial solution given by

(14)
$$S_{k+1}^+ = S_k^+ \cup J^k$$

$$S_{k+1}^- = S_k^- \, ,$$

which corresponds to x^k. Since (14) corresponds to a feasible CP solution, backtracking occurs immediately.

Computational Results

Some typical runs on the IBM 360/50 are

m	n	density	time (min.)
30	90	.07	.6
50	450	.046	2.15

8. AN ENUMERATION ALGORITHM FOR THE PP [25,26]

For the PP, a very specialized implicit enumeration algorithm has performed very well. Extensive use is made of the results of Section 4 on handling binary data.

Initially every column of A is placed in one of m lists. Column j is placed in list k if $k = \min\{i \,|\, a_{ij} = 1\}$. Note that some lists may be empty. However, if list 1 is empty, there is no feasible solution. Within each list, the columns are partially ordered by the cost coefficients. In particular,

$$k, t \,\epsilon\, \text{list } i, \quad k \text{ precedes } t \Rightarrow c_k \leq c_t \, .$$

It is convenient to introduce notation slightly different from that of Section 7. Let S denote a partial solution, $S^+ = \{j \,|\, x_j = 1, j \,\epsilon\, S\}$ and $z(S) = \sum_{j \in S^+} c_j$. S will always be chosen such that $\sum_{j \in S^+} a_{ij} \leq 1$, for all i. Let the constraints satisfied by S be

$$Q(S) = \bigcup_{j \in S^+} P_j \, .$$

Algorithm

Step 1 (Initialization). Let $S = \phi$, $\bar{z} = \infty$. Go to Step 2.

Step 2 (Choose next list). Let $i* = \min\{i \mid i \notin Q(S)\}$. Set an indicator at the top (lowest cost element) of list $i*$. Go to Step 3.

Step 3 (Test for augmenting variable). Begin at the indicated position in list $i*$. Examine the columns of the list in order. If a column j is found such that $Q(S) \cap P_j = \phi$ and $z(S) + c_j < \bar{z}$ go to Step 4. If a column j is reached such that $z(S) + c_j \geq \bar{z}$, or if list $i*$ is exhausted, go to Step 5.

Step 4 (Test for solution). Let $S^+ = S^+ \cup \{j\}$. If $Q(S) = I$ a better solution has been found, let $\bar{z} = z(S)$ and go to Step 5. If $Q(S) \neq I$ go to Step 2.

Step 5 (Backtrack). If $S^+ = \phi$ terminate. If $\bar{z} < \infty$ an optimal solution has been found, otherwise none exists. If $S^+ \neq \phi$, let $\{k\}$ be the last element included in S^+. Let $S^+ = S^+ - \{k\}$. Let $i*$ be the list in which k is found, and set an indicator directly below k in list $i*$. Go to Step 3.

Although this algorithm is related to the enumeration scheme of Section 7, an important difference occurs in Step 3 where the search for a branching variable and the possibility of fathoming can be limited to examining list $i*$. This search takes advantage of the simple fact that if $x_j = 1$, every list $i \in P_j$ can be skipped. Thus it is not necessary to solve any LP's in the branching and fathoming tests. In fact, frequently a PP can be solved more quickly than the corresponding PP', which is very unusual for integer programming problems.

Since every solution contains an element of list 1, the search actually terminates when list 1 is exhausted. For this reason, it seems logical to have list 1 be as small as possible. This may be achieved by permuting the rows so that $\sum_j a_{1j} = \min_i \sum_j a_{ij}$.

It is extremely important for efficient computation that the basic test in the algorithm $Q(S) \cap P_j \overset{?}{=} \phi$ be implemented by using the logical AND operation described in Section 4. Thus the algorithm should be coded in an assembly language.

A simple surrogate constraint has been employed in the algorithm just given with mixed success [27]. At any partial solution S, the resulting PP is

$$\min \ z_k \ = \ \sum_{j \notin S} c_j x_j \ + \ z(S)$$

(15)
$$\sum_{j \notin S} a_{ij} x_j \ = \ 1 \qquad i \notin Q(S)$$

$$x_j \ = \ 0,1 \qquad j \notin S \ .$$

Adding the constraints of (15) yields the problem

$$\min \ \sum_{j \notin S} c_j x_j \ + \ z(S)$$

(16)
$$\sum_{j \notin S} v_j x_j \ = \ t$$

$$x_j \ = \ 0,1 \qquad j \notin S$$

where $v_j = \sum\limits_{i \notin Q(S)} a_{ij}$ and $t = m - |Q(S)|$. Problem (16) is a 0-1
knapsack problem and could be solved to yield a lower bound on z_k.
A weaker bound which is more easily obtained is

$$\underline{z}(S) \ = \ t c_{j*}/v_{j*} \ + \ z(S)$$

where $c_{j*}/v_{j*} = \min\limits_{j \notin S} c_j/v_j$. Backtracking occurs if $<\underline{z}(S)> \ \geq \ \bar{z}$.

Computational Results [26]

This algorithm has been programmed in Fortran and MAP (IBM 7094 assembly language). Some typical results are

m	n	density	time (sec.)
37	200	.3	5
37	200	.1	257
26	777	.16	39

In general the algorithm runs faster for high density problems since list skipping occurs more often. Results in [25] and [27] are comparable.

9.   COVERING AND MATCHING ON GRAPHS

Although the theory and methodology given in Sections 5-7 apply to the GP and MP, much stronger results have been obtained for this class of problems. To describe these results, it is necessary to introduce some new notation and terminology.

Given an undirected graph $G = (V,E)$ and a subset $S \subseteq E$, define the degree of vertex i with respect to S as

$$d_S(i) = \text{number of edges of } S \text{ incident to } i.$$

A subset $M \subseteq E$ is called a <u>matching</u> of G if $d_M(i) \leq 1$ for all $i \in V$, and a subset $C \subseteq E$ is called a <u>cover</u> of G if $d_C(i) \geq 1$ for all $i \in V$. Every graph trivially contains a matching since $d_\phi(i) = 0$, for all $i \in V$. Every connected graph contains a cover since $d_E(i) \geq 1$ for all $i \in V$. Without loss of generality it can be assumed that G is connected; if it is not, all of the results apply separately to each component of G.

Recall that in the MP, the objective is to maximize $w(M) = \sum_{e_j \in M} c_j$ and in the GP to minimize $w(C) = \sum_{e_j \in C} c_j$. If $c_j = 1$ for all j, the MP problem is to find a matching of maximum cardinality and the GP is to find a cover of minimum cardinality. These two problems have been called the simple matching problem (SMP) and the simple covering problem (SGP), respectively.

Finding an optimal cover to an SGP essentially solves the SMP over the same graph and conversely since

<u>Theorem 7</u> [28] Let C^* be an optimal solution to the SGP over G. For every vertex i having $d_{C^*}(i) > 1$, remove $d_{C^*}(i)-1$ of the edges incident to it. The resulting set of edges is an optimal solution to the SMP over G.

<u>Theorem 8</u> [28] Let M* be an optimal solution to the SMP over G.
Add one edge incident to every vertex such that $d_{M*}(i) = 0$. The
resulting set of edges is an optimal solution to the SGP over G.

　　　For the remainder of this section only the MP will be considered; all of the results have analogues for the GP.

Matching on a Bipartite Graph

　　　A graph G = (V,E) is called <u>bipartite</u> if there exist sets
V_1 and V_2 such that

$$V_1 \cup V_2 = V, \quad V_1 \cap V_2 = \phi$$

and every edge of G is incident to one vertex of V_1 and one vertex of V_2. For bipartite graphs the matrix A of (4) is totally
unimodular, which implies that such an MP can be solved by linear
programming. But even more strongly, an MP on a bipartite graph
is an assignment problem.

　　　To construct the assignment problem, let V_1 = {set of men}
and V_2 = {set of jobs}. If $|V_1| \neq |V_2|$ add dummy vertices to
the smaller set to achieve equality with respect to the cardinality
of the sets. For $i \in V_1$ and $j \in V_2$ define

$$c_{ij} = \begin{cases} c_k & \text{if there is an edge } e_k \text{ in } G \text{ joining} \\ & \text{i and j} \\ -M & \text{if there is no such edge, or if either} \\ & \text{i or j is a dummy vertex, where M is} \\ & \text{a very large positive number} \end{cases}$$

and let c_{ij} be the value of assigning man i to job j. It is
then easy to see that a maximum value assignment can be interpreted
as a maximum matching on G. In particular denote an optimal solution to the assignment problem by x_{ij}^o, where

$$x_{ij}^o = \begin{cases} 1 & \text{if man } i \text{ is assigned to job } j \\ 0 & \text{otherwise .} \end{cases}$$

Then an optimal solution to the corresponding MP is given by

$$E^o = \{e_k | e_k = (i,j), \quad x_{ij}^o = 1, \quad (i,j) \in E\}.$$

The Convex Hull of Solutions to an MP

Since a graph is bipartite if and only if it contains no odd cycles (cycles with an odd number of edges), the above discussion indicates that the difficulty in solving an MP is caused by odd cycles in G. Specifically, consider an SMP for which the edge sequence $(e_1, e_2, \ldots, e_{2k+1})$ is an odd cycle. Let the vertex sequence of this cycle be given by $(i_1, \ldots, i_{2k+1}, i_1)$. Then the constraints $\sum_{j=1}^{2k+1} a_{ij} x_j \leq 1$, $i = i_1, \ldots, i_{2k+1}$, where $A = \{a_{ij}\}$ is the vertex-edge incidence matrix of G, are satisfied by $x_1^o = \cdots = x_{2k+1}^o = 1/2$ and yield

$$\sum_{j=1}^{2k+1} x_j^o = k + 1/2 .$$

However, in a matching, no more than k edges from an odd cycle of 2k+1 edges can be used. Thus when a matching problem is written as an integer program, the constraint

(17)
$$\sum_{j=1}^{2k+1} x_j \leq k$$

where $(e_1, \ldots, e_{2k+1})$ is an odd cycle, is a valid cutting plane. But constraints of the form (17) accomplish considerably more, since

Theorem 9 [29,30,31] The convex hull of solutions to $\sum_j a_{ij} x_j \leq 1$, $i = 1, \ldots, m$, $x_j = 0, 1$ $j = 1, \ldots, n$, where A is the incidence matrix of G is given by

$$\sum_j a_{ij}x_j \leq 1, \qquad i = 1,\ldots,m$$

$$(18) \qquad \sum_{j=1}^{2k+1} x_j \leq k, \qquad \text{for every odd cycle containing 2k+1}$$
$$\text{edges,} \quad k = 1,2,\ldots$$

$$x_j \geq 0, \qquad j = 1,\ldots,n$$

Matching problems belong to a very small class of integer programming problems for which the convex hull of solutions has been explicitly identified. Note that Theorem 9 states that an MP can be solved as an LP with constraint set (18) or by a cutting plane algorithm which would add a constraint (17) whenever an optimal LP solution was obtained which did not satisfy $x_j = 0,1$. However this approach is not likely to be as efficient as the graph-oriented algorithm described below.

Augmenting Paths

Relative to a set $S \subseteq E$, an <u>alternating path</u> in G is a simple path whose edges alternate between S and E-S. Relative to a matching M, an <u>augmenting path</u> is an alternating path connecting vertices i and k such that $d_M(i) = d_M(k) = 0$. Note that an augmenting path must contain an odd number of edges. Figure 1 shows a matching $M = \{e_2\}$, an augmenting path $p = (e_1,e_2,e_3)$ and an improved matching $M' = \{e_1,e_3\}$ to an SMP.

Figure 1

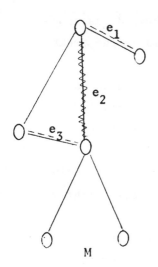

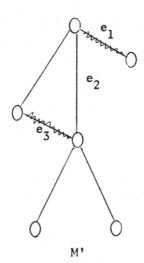

Clearly, if an augmenting path exists, a better matching to an SMP can be obtained by reversing the assignments along the augmenting path. But even more strongly,

Theorem 10 [28],[32]

A matching M is maximum to an SMP if and only if there is no augmenting path relative to M.

Theorem 10 yields the following algorithm for an SMP [28].

Step 1 - Begin with an arbitrary matching M. (Possibly ϕ).

Step 2 - From each vertex i having $d_M(i) = 0$ search in turn for an augmenting path. If one is found, go to Step 3. If none is found, the matching is maximum.

Step 3 - Reverse the assignments along the augmenting path and go to Step 2.

Certainly the algorithm given above will work. Its drawback is in Step 2. A "good" algorithm must have an efficient way to search for augmenting paths. In [32] and [33] such

algorithms are developed with upper bounds on the number of calcu-
lations on the order of $|E|^2$. A modified version of the algorithm
of [32] is given in [34].

To explain and justify these algorithms is beyond the scope
of this paper. However, the key to both algorithms is the treatment of odd cycles in the search for an augmenting path. The
method of [32] involves reducing the graph by shrinking an odd
cycle into a single artificial vertex, when one is identified.
The method of [33] always operates on the original graph and
keeps track of odd cycles by a set of vertex labeling rules.

An empirical comparison of the two algorithms has recently
been carried out in [35]. The results of [35] indicate slightly
faster running times for the algorithm given in [33]. The
largest problem considered in [35] is for a graph with 60
vertices and 263 edges. It was solved in less than 2 seconds
on an IBM 360/50 with both methods.

Theorem 10 has been extended to MP's using a generalized
definition of an augmenting path [33]. Define an augmenting
path p, relative to M, to be an alternating path or alternating cycle having no edge of M incident to exactly one vertex
of p, and with the property

$$\sum_{e_j \epsilon P-M} c_j - \sum_{e_j \epsilon P \cap M} c_j > 0$$

where P is the set of edges contained in the path p. Using
this definition of an augmenting path, Theorem 10 applies to
MP's.

In [29], the algorithm of [32] is extended to MP's. A
further extension is made in [36] to problems in which $\sum_j a_{ij} \leq 1$
is generalized to $\sum_j a_{ij} \leq b_i$, where b_i is a positive integer
and the condition $a_{hj} = a_{kj} = 1$, $a_{ij} = 0$ otherwise is gener-
alized to $\sum_i |a_{ij}| \leq 2$, a_{ij} integer.

Here $a_{ij} = -1$ permits directed edges and $a_{ij} = 2$ allows edges of the form $e_j = (i,i)$. It is indicated in [36] that this general algorithm has been programmed and problems with 300 vertices, 1500 edges $b_i = 1,2$ and $1 \le c_j \le 10$ take about 30 seconds on a high-speed digital computer.

References

[1]   R. Garfinkel, "Set Covering: A Survey," presented at the
      17th International Conference of the Institute of Manage-
      ment Sciences, July, 1970.

[2]   C. E. Lemke, H. M. Salkin, and K. Spielberg, "Set Covering
      by Single Branch Enumeration with Linear Programming Sub-
      problems," Operations Research 19, 998-1022 (1971).

[3]   R. H. Day, "On Optimal Extracting From a Multiple File Data
      Storage System: An Application of Integer Programming,"
      Operations Research 13, 482-494 (1965).

[4]   M. Bellmore, H. J. Greenberg, and J. J. Jarvis, "Multi-
      Commodity Disconnecting Sets," Management Science 16,
      B427-433 (1970).

[5]   M. L. Balinski, and R. Quandt, "On an Integer Program for a
      Delivery Problem," Operations Research 12, 300-304 (1964).

[6]   W. H. Wagner, "An Application of Integer Programming to
      Legislative Redistricting," presented at the 34th National
      Meeting of ORSA, November, 1968.

[7]   R. S. Garfinkel, and G. L. Nemhauser, "Optimal Political
      Districting by Implicit Enumeration Techniques," Manage-
      ment Science 16, B495-508 (1970).

[8]   T. N. Kolner, "Some Highlights of a Scheduling Matrix
      Generator System," United Airlines, presented at the Sixth
      AGIFORS Symposium, Sept. 1966.

[9]   J. P. Arabeyre, J. Fearnley, F. C. Steiger, and W. Teather,
      "The Airline Crew Scheduling Problem: A Survey," Transp.
      Sci. 3, 140-163 (1969).

[10] F. Bessiere, "Sur la Recherche du Nombre Chromatique d'Un
      Graphe par un Programme Lineaire en Nombres Entiers,"
      Rev. Franc. Recherche Operat. 9, 143-148 (1965).

[11] A. Cobham, R. Fridshal, and J. H. North, "An Application of
      Linear Programming to the Minimization of Boolean Functions,"
      Research Report RC-472, IBM Research Center, June, 1961.

[12] M. C. Paul, and S. H. Unger, "Minimizing the Number of States
      in Incompletely Specified Sequential Functions," IRE Trans-
      actions on Electronic Computers EC-8, 356-367 (1959).

[13] M. Bellmore, and H. D. Ratliff, "Optimal Defense of Multi-Commodity Networks," The Johns Hopkins University, 1969.

[14] G. B. Dantzig, and J. H. Ramser, "The Truck Dispatching Problem," Management Science 6, 80-91 (1959).

[15] G. Clarke, and S. W. Wright, "Scheduling of Vehicles from a Central Depot to a Number of Delivery Points," Operations Research 12, 568-581 (1964).

[16] M. E. Salveson, "The Assembly Line Balancing Problem," Journal of Industrial Engineering 6, 18-25 (1955).

[17] M. Bellmore, and H. D. Ratliff, "Set Covering and Involutory Bases," The Johns Hopkins University, Baltimore, Maryland, 1969.

[18] G. Andrew, T. Hoffman, and C. Krabek, "On the Generalized Set Covering Problem," Control Data Corporation, presented at the ORSA-TIMS Conference, May, 1968.

[19] E. Balas, and M. W. Padberg, "On the Set Covering Problem," Management Science Research Report No. 197, Carnegie-Mellon University, February, 1970.

[20] R. W. House, L. D. Nelson, and J. Rado, "Computer Studies of a Certain Class of Linear Integer Problems," in Recent Advances in Optimization Techniques, edited by A. Lavi and T. Vogl, Wiley, New York, 1966.

[21] A. M. Geoffrion, "An Improved Implicit Enumeration Approach to Integer Programming," Operations Research 17, 437-454 (1969).

[22] E. Balas, "An Additive Algorithm for Solving Linear Programs with 0-1 Variables," Operations Research 13, 517-546 (1965).

[23] A. M. Geoffrion, "Integer Programming by Implicit Enumeration and Balas' Method," SIAM Rev. 7, 178-190 (1967).

[24] E. Balas, "Discrete Programming by the Filter Method," Operations Research 15, 915-957 (1967).

[25] J. F. Pierce, "Application of Combinatorial Programming to a Class of All-Zero-One Integer Programming Problems," Management Science 15, 191-209 (1968).

[26] R. S. Garfinkel, and G. L. Nemhauser, "The Set Partitioning Problem: Set Covering with Equality Constraints," Operations Research 17, 848-856 (1969).

[27] J. F. Pierce, and S. Lasky, "Improved Combinatorial Program-
      ming Algorithms for a Class of All-Zero-One Integer Program-
      ming Problems," IBM Cambridge Scientific Center Report,
      June, 1970.

[28] R. Z. Norman and M. O. Rabin, "An Algorithm for the Minimum
      Cover of a Graph," Proc. Am. Math. Soc. 10, 315-319 (1959).

[29] J. Edmonds, "Maximum Matching and a Polyhedron with 0,1-
      Vertices," J. of Res. of the Nat. Bur. Stds. 69B, 125-130
      (1965).

[30] M. L. Balinski, "Establishing the Matching Polytope," City
      Univ. of New York, 1969.

[31] M. W. Padberg, "Simple Zero-One Problems: Set Coverings,
      Matchings and Covering in Graphs," Management Science
      Research Report No. 235, Carnegie-Mellon Univ., 1971.

[32] J. Edmonds, "Paths, Trees and Flowers," Can. J. Math. 17,
      449-467 (1965).

[33] M. L. Balinski, "Labelling to Obtain a Maximum Matching,"
      Proc. Conf. on Comb. Math. and Its Appl. - 1967, Univer-
      sity of North Carolina (to appear).

[34] C. Witzgall and C. T. Zahn, Jr., "Modification of Edmonds'
      Maximum Matching Algorithm," J. of Res. of the Nat. Bur. Stds
      69B, 91-98 (1965).

[35] B. F. Gordon, "The Maximum Matching Problem - A Comparison of
      the Edmonds and Balinski Algorithms," Grad. School of Manage-
      ment, Univ. of Rochester, 1971.

[36] J. Edmonds and E. L. Johnson, "Matching: A Well Solved Class
      of Integer Linear Programs," Proc. of the Calgary Int. Conf.
      on Comb. Structures and Their Appl. 89-92, Gordon and Breach
      (1970).

[37] M. L. Balinski, "On Maximum Matching, Minimum Covering and
      their Connections," Proc. Princeton Sym. on Math. Prog. -
      1967, (H. W. Kuhn, ed.) 303-312 (1970).

[38] R. Roth, "Computer Solutions to Minimum Cover Problems,"
      Operations Research 17, 455-466 (1969).

[39] E. L. Lawler, "Covering Problems: Duality Relations and a
      New Method of Solution," SIAM J. Appl. Math. 14, 1115-1132
      (1966).

[40] P. A. Jensen, "Optimum Network Partitioning," Operations
      Research 19, 916-932 (1971).

[41] F. Glover, "Maximum Matching in a Convex Bipartite Graph,"
      Naval Res. Log. Quart. 14, 313-316 (1967).

[42] J. F. Desler and S. L. Hakimi, "A Graph-Theoretic Approach
      to a Class of Integer-Programming Problems," Operations
      Research 17, 1017-1033 (1969).

[43] J. Edmonds, "Covers and Packings in a Family of Sets," Bull.
      Am. Math. Soc. 68, 494-499 (1962).

[44] L. J. White, "Minimum Covers of Fixed Cardinality in Weighted
      Graphs," SIAM J. 21, 104-113 (1971).

[45] D. R. Morrison, "Matching Algorithms," J. of Combinatorial
      Theory 6, 20-32 (1969).

[46] D. de Werra, "On Some Combinatorial Problems Arising in
      Scheduling," CORS 8, 165-175 (1970).

[47] R. E. Marsten, "An Algorithm for the Set Partitioning Problem
      with Side Constraints," Ph.D. Thesis, UCLA, 1971.

IV. OPTIMIZATION IN NETWORKS

7. FLOW NETWORKS AND COMBINATORIAL OPERATIONS RESEARCH

D. R. FULKERSON, RAND, Santa Monica, California

PART I: FLOWS IN NETWORKS

1. Maximal flow. A *directed network* (graph) $G = [N; \mathcal{Q}]$ consists of a finite collection N of elements 1, 2, $\cdots$, n together with a subset $\mathcal{Q}$ of the ordered pairs (i, j) of distinct elements of N. The elements of N will be called *nodes*; members of $\mathcal{Q}$ are *arcs*. Figure 1.1 shows a directed network having four nodes and six arcs (1, 2), (1, 3), (2, 3), (2, 4), (3, 2), and (3, 4).

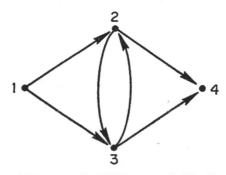

FIG. 1.1

Sometimes we shall also consider *undirected networks*, for which the set $\mathcal{Q}$ consists of unordered pairs of nodes. For emphasis, these will then be termed *links*.

Suppose that each arc (i, j) of a directed network has associated with it a nonnegative number c_{ij}, the *capacity* of (i, j), to be thought of as representing the maximal amount of some commodity that can arrive at j from i along (i, j) per unit time in a steady-state situation. Then a natural question is: What is the maximal amount of commodity-flow from some node to another via the entire network? (For example, one might think of a network of city streets, the commodity being cars, and ask for a maximal traffic flow from some point to another.) We may formulate the question mathematically as follows. Let 1 and n be the two nodes in question. A *flow, of amount v,* from 1 *to n* in $G = [N; \mathcal{Q}]$ is a function x from $\mathcal{Q}$ to real numbers (a vector x having components x_{ij} for (i, j) in $\mathcal{Q}$) that satisfies the linear equations and inequalities

$$(1.1) \qquad \sum_j x_{ij} - \sum_j x_{ji} = \begin{cases} v, & i = 1, \\ -v, & i = n, \\ 0, & \text{otherwise,} \end{cases}$$

$$(1.2) \qquad 0 \leq x_{ij} \leq c_{ij}, \qquad (i, j) \text{ in } \mathcal{Q}.$$

In (1.1) the sums are of course over those nodes for which x is defined. We call 1 the *source*, n the *sink*. A *maximal flow* from source to sink is one that maximizes the variable v subject to (1.1), (1.2).

Figure 1.2 shows a flow from source node 1 to sink node 6 of amount 7. In Figure 1.2, the first number of each pair beside an arc is the arc capacity, the second number the arc flow.

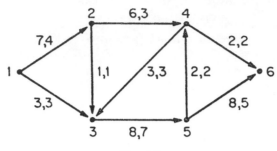

FIG. 1.2

To state the fundamental theorem about maximal flow, we need one other notion, that of a cut. A *cut separating* 1 and n is a partition of the nodes into two complementary sets, I and J, with 1 in I, say, and n in J. The *capacity* of the cut is then

(1.3)
$$\sum_{\substack{i \text{ in } I \\ j \text{ in } J}} c_{ij}.$$

(For instance, if $I = \{1, 3, 4\}$ in Fig. 1.2, the cut has capacity $c_{12}+c_{35}+c_{46}=17$.) A cut separating source and sink of minimum capacity is a *minimal* cut, relative to the given source and sink.

Summing the equations (1.1) over i in the source-set I of a cut and using (1.2) shows that

(1.4)
$$v = \sum_{\substack{i \text{ in } I \\ j \text{ in } J}} (x_{ij} - x_{ji}) \leqq \sum_{\substack{i \text{ in } I \\ j \text{ in } J}} c_{ij}.$$

In words, for an arbitrary flow and arbitrary cut, the net flow across the cut is the flow amount v, which is consequently bounded above by the cut capacity. Theorem 1.1 below asserts that equality holds in (1.4) for some flow and some cut, and hence the flow is maximal, the cut minimal [11].

THEOREM 1.1. *For any network the maximal flow amount from source to sink is equal to the minimal cut capacity relative to the source and sink.*

Theorem 1.1 is a kind of combinatorial counterpart, for the special case of the maximal flow problem, of the duality theorem for linear programs, and can be deduced from it [5]. But the most revealing proof of Theorem 1.1 uses a simple "marking" or "labeling" process [12] for constructing a maximal flow, which also yields the following theorem.

THEOREM 1.2. *A flow x from source to sink is maximal if and only if there is no flow-augmenting path with respect to x.*

Here we need to say what an x-augmenting path is. First of all, a path from one node to another is a sequence of distinct end-to-end arcs that starts at the first node and terminates at the second; arcs traversed with their direction in going along the path are *forward* arcs of the path, while arcs traversed against their direction are *reverse* arcs of the path. A path from source to sink is x-augmenting provided that $x < c$ on forward arcs and $x > 0$ on reverse arcs. For example, the path (1, 2), (2, 4), (5, 4), (5, 6) in Figure 1.2 is an augmenting path for the flow shown there. Figure 1.3 below indicates how such a path can be used to increase the amount of flow from source to sink.

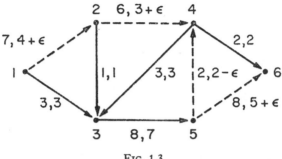

FIG. 1.3

Taking the flow change ϵ along the path as large as possible in Figure 1.3, namely $\epsilon = 2$, produces a maximal flow, since the cut $I = \{1, 2, 4\}$, $J = \{3, 5, 6\}$ is then "saturated."

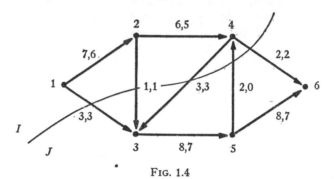

FIG. 1.4

The labeling process of [12] is a systematic and efficient search, fanning out from the source, for a flow augmenting path. If none such exists, the process ends by locating a minimal cut.

The following theorem, of special significance for combinatorial applications, is also a consequence of the procedure sketched above for constructing maximal flow.

THEOREM 1.3. *If all arc capacities are integers, there is an integral maximal flow.*

It is sometimes convenient to alter the constraints (1.2) of the maximal flow problem to

(1.5) $l_{ij} \leq x_{ij} \leq c_{ij}.$

Here l is a given lower bound function satisfying $l \leq c$. The analogue of Theorem 1.1 is then

THEOREM 1.4. *If there is a function x satisfying (1.1) and (1.5) for some number v, then the maximum v subject to these constraints is equal to the minimum of*

(1.6) $$\sum_{\substack{i \text{ in } I \\ j \text{ in } J}} (c_{ij} - l_{ji})$$

taken over all cuts I, J separating source and sink. On the other hand, the minimum v is equal to the maximum of

(1.7) $$\sum_{\substack{i \text{ in } I \\ j \text{ in } J}} (l_{ij} - c_{ji})$$

taken over all cuts I, J separating source and sink.

The question of the existence of such a flow x, together with another flow feasibility question, will be discussed in the next section.

2. Feasibility theorems. The constraints of the maximal flow problem are of course always feasible, since $x = 0$ satisfies (1.1), (1.2) with $v = 0$. By changing the constraints in various ways, interesting feasibility questions arise. Here we shall consider two such, one involving supplies and demands at nodes, the other lower bounds on arc flows, as in (1.5).

Let $G = [N; \mathcal{C}]$ have capacity function c, and let S and T be disjoint subsets of N. With each i in S associate a *supply* $a_i \geq 0$, with each i in T a *demand* $b_i \geq 0$, and impose the constraints

(2.1) $$\sum_j (x_{ij} - x_{ji}) \leq a_i, \quad i \text{ in } S,$$
$$\leq -b_i, \ i \text{ in } T,$$
$$= 0, \qquad \text{otherwise,}$$

(2.2) $0 \leq x_{ij} \leq c_{ij}, \quad (i, j) \text{ in } \mathcal{C}.$

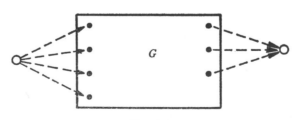

FIG. 2.1

In words, the net flow out of i in S is bounded above by the supply a_i, and the net flow into i in T is bounded below by the demand b_i. When are the supply-demand constraints (2.1), (2.2) feasible?

This question is easily answered by applying Theorem 1.1 to an enlarged network. Extend $G = [N; \mathcal{a}]$ to $G^* = [N^*; \mathcal{a}^*]$ by adjoining a source 0 and sink $n+1$, together with source arcs $(0, j)$ for j in S, and sink arcs $(i, n+1)$ for i in T. (See Figure 2.1.) The capacity function c^* on $\mathcal{a}^*$ is defined by $c^*_{0,j} = a_j$ for j in S, $c^*_{i,n+1} = b_i$ for i in T, $c^*_{ij} = c_{ij}$ for (i, j) in $\mathcal{a}$. The constraints (2.1) and (2.2) are feasible if and only if the maximal flow amount from source to sink in the enlarged network is at least $\sum_{i \text{ in } T} b_i$, that is, if and only if a maximal flow saturates all sink arcs. Hence we need only construct a maximal flow in order to check the feasibility of (2.1), (2.2). By pushing the analysis a little further, using Theorem 1.1, the following theorem emerges [21].

THEOREM 2.1. *The supply-demand constraints (2.1), (2.2) are feasible if and only if, for each subset T' of T, there is a flow $x(T')$ that satisfies the aggregate demand $\sum_{i \text{ in } T'} b_i$ without violating the supply limitations at nodes of S.*

Here satisfying the aggregate demand over T' means that the net flow into the set T' must be at least $\sum_{i \text{ in } T'} b_i$, without regard for the individual demands in T'. The necessity of the condition is of course clear; sufficiency asserts the existence of a single flow x meeting all individual demands provided the flows $x(T')$ exist for all subsets T' of T.

It should be noted that if the functions a, b, c of (2.1), (2.2) are integral valued, and if feasible flows exist, then there is an integral feasible flow. This follows from Theorem 1.3 and the conversion of (2.1), (2.2) to a maximal flow problem. A similar integrity statement holds for the situation of Theorem 1.4, and indeed, for all the flow problems to be discussed in any detail in this survey.

We turn now to a consideration of lower bounds on arc flows, as in (1.5), and pose the resulting feasibility question in terms of circulations, i.e., flows that are source-sink free, instead of flows from source to sink. (One can always add a "return-flow" arc from sink to source to convert to circulations.) Thus we are questioning the feasibility of the constraints

$$(2.3) \qquad \sum_j (x_{ij} - x_{ji}) = 0, \qquad i \text{ in } N,$$

$$(2.4) \qquad l_{ij} \leqq x_{ij} \leqq c_{ij}, \qquad (i, j) \text{ in } \mathcal{a}.$$

The following theorem answers the question [26]. Its proof can be made to rely on Theorem 1.1 [15].

THEOREM 2.2. *The constraints (2.3), (2.4) are feasible if and only if*

$$(2.5) \qquad \sum_{\substack{i \text{ in } I \\ j \text{ in } J}} c_{ij} \geqq \sum_{\substack{i \text{ in } I \\ j \text{ in } J}} l_{ji}$$

holds for all partitions I, J of N.

Again the necessity is clear, since (2.5) simply says there must be sufficient escape capacity from the set I to take care of the flow forced into I by the function l. But sufficiency is not obvious.

Other useful flow feasibility theorems have been deduced [19, 26]. In each case Theorem 1.1 can be used as the main tool in a proof.

3. Minimal cost flows. One of the most practical problem areas involving network flows is that of constructing flows satisfying constraints of various kinds and minimizing cost. The standard linear programming transportation problem, which has an extensive literature, is in this category.

We put the problem as follows. Each arc (i, j) of a network $G = [N; \alpha]$ has a capacity c_{ij} and a cost a_{ij}. It is desired to construct a flow x from source to sink of specified amount v that minimizes the total flow cost

$$(3.1) \qquad \sum_{(i,j) \text{ in } \alpha} a_{ij} x_{ij}$$

over all flows that send v units from source to sink. In many applications one has supplies of a commodity at certain points in a transportation network, demands at others, and the objective is to satisfy the demands from the supplies at minimum cost.

By treating v as a parameter, the method for constructing maximal flows can be used to construct minimal cost flows throughout the feasible range of v. Indeed, the solution procedure can be viewed as one of solving a sequence of maximal flow problems, each on a subnetwork of the original one [14]. Another, not essentially different, viewpoint is provided by the following theorem [1, 29].

THEOREM 3.1. *Let x be a minimal cost flow from source to sink of amount v. Then the flow obtained from x by adding $\epsilon > 0$ to the flow in forward arcs of a minimal cost x-augmenting path, and subtracting ϵ from the flow in reverse arcs of this path, is a minimal cost flow of amount $v + \epsilon$.*

Here the cost of a path is the sum of arc costs over forward arcs minus the corresponding sum over reverse arcs, i.e., the cost of "sending an additional unit" via the path.

Thus, if all arc costs a_{ij} are nonnegative, for example, one can start with the zero flow and apply Theorem 3.1 to obtain minimal cost flows for increasing v. (The cost profile thereby generated is piecewise linear and convex.) All that is needed to make this an explicit algorithm is a method of searching for a minimal cost flow augmenting path. Various ways of doing this can be described. One such will be given in Part II, section 1.

Another method [17] for constructing minimal cost flows poses the problem in circulation format, that is, (3.1) is to be minimized subject to (2.3), (2.4). This construction has a number of advantages, principally in terms of generality and flexibility. For instance, it may be started with any circulation; even (2.4) need not be satisfied initially. Also, no assumption about the cost function is required.

These methods produce integral flows in case the arc capacities (and lower bounds) are integers. Theoretical upper bounds on the computing task, ones that are quite good, are easily obtained in each case. This may be contrasted with the situation for general linear programs, where decent upper bounds on solution methods are unknown.

4. Maximal dynamic flow. Suppose that each arc (i, j) of a network G has not only a capacity, but a transit time t_{ij} as well, and that we are interested in determining the maximum amount of flow that can reach sink n from source 1 in a specified number t of time periods. This dynamic flow problem can always be treated as a static flow problem in a time-expanded version G_t of G. For example, if the given network G is that of Figure 4.1 and if each arc of G has unit transit time, then G_3 is shown in Figure 4.2. (We have included "storage arcs" leading from a location to itself one unit of time later.)

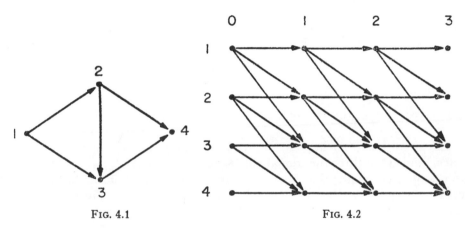

FIG. 4.1   FIG. 4.2

Expanding the network in this way puts one back in the static case. Moreover, arc capacities and transit times can vary with time and this is still so. However, if each capacity and transit time is fixed over time, the problem can be solved in the smaller network G. Specifically, a maximal dynamic flow for t periods can be generated from a static flow x in G of amount v that minimizes the linear form

$$(4.1) \qquad \sum_{(i,j) \text{ in } \mathcal{C}} t_{ij}x_{ij} - (t + 1)v$$

over all flows in G from source 1 to sink n [14]. By adding the return-flow arc $(n, 1)$ to G with $c_{n1} = \infty$, $t_{n1} = -(t+1)$, the problem may be viewed as one of constructing a circulation that minimizes the "cost" form (4.1).

5. Multi-terminal maximal flow. Heretofore we have phrased statements in terms of directed networks. In this section we confine the discussion to undirected flow networks, by which we merely mean the following. A link (i, j) can carry flow in either direction and has the same flow capacity each way. Thus one can think of an arc (i, j) with capacity c_{ij} and an arc (j, i) with capacity $c_{ji} = c_{ij}$. The assumption of a symmetric capacity function c makes the results described in this section considerably simpler and more appealing than they would otherwise be.

Instead of dealing with a single source and sink, we shift attention to all pairs of distinct nodes taken as terminals for flows. These flows are not to be thought of as occurring simultaneously.

Let v_{ij} denote the maximal flow amount from i to j. Thus the function v is symmetric, $v_{ij} = v_{ji}$, and may be determined explicitly for an n-node network by solving $n(n-1)/2$ maximal flow problems. There is, however, a much simpler way of determining the function v, one that involves the solution of only $n-1$ maximal flow problems; in addition, there is a simple condition in order that a symmetric function v be realizable as the maximal multiterminal flow function of some undirected network [22].

THEOREM 5.1. *A symmetric, nonnegative function v is realizable by an undirected network if and only if v satisfies*

$$(5.1) \qquad\qquad v_{ij} \geqq \min (v_{ik}, v_{kj})$$

for all triples i, j, k.

The necessity of the "triangle inequality" (5.1) follows easily from Theorem 1.1.

The condition (5.1) imposes severe limitations on the function v. For instance, among the three functional values appearing in (5.1), two must be equal and the third no smaller than their common value. It also follows that if the network has n nodes, v can take on at most $n-1$ numerically different functional values. It is not altogether surprising, therefore, that v can be determined by a

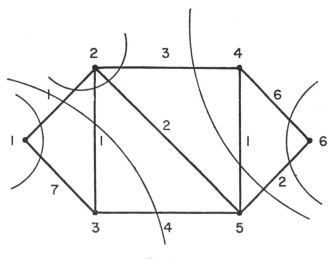

FIG. 5.1

simpler process than solving all single-terminal maximal flow problems. This process systematically picks out precisely $n-1$ cuts in the network having the property that v_{ij} is determined by the minimum one of these cuts separating i and j [22]. For example, in the network of Figure 5.1 the relevant cuts are those shown. Thus, for instance, since nodes 1 and 4 are separated by the three cuts $(1/2, 3, 4, 5, 6)$, $(1, 3/2, 4, 5, 6)$, $(1, 2, 3, 5/4, 6)$ having capacities 8, 6, 6 respectively, then $v_{14} = \min(8, 6, 6) = 6$.

6. Other flow problems. The flow problems that have been discussed thus far all have the useful and pleasant feature that the assumption of integral data implies the existence of an integral solution. A number of flow problems that do not share this property have also been studied. Among these we mention flows in networks with gains [29], simultaneous multi-terminal flows [13], and problems involving optimal synthesis of flow networks that meet specified requirements [16, 22]. Methods of solution for such problems are, in general, more complicated than methods for those we have discussed. One rather surprising exception to this statement is the following synthesis problem. Suppose it is desired to construct an undirected network on a specified number of nodes so that $v_{ij} \geq r_{ij}$ for stipulated requirements r_{ij}, with the total sum of link capacities of the network minimal. A very simple combinatorial method of solution for this synthesis problem is given in [22].

PART II: COMBINATORIAL PROBLEMS

1. Network potentials and shortest chains. Consider a directed network in which each arc (i, j) has associated with it a positive number a_{ij}, which may be thought of as the length of the arc, or the cost of traversing the arc. How does one determine a shortest chain from some node to another? Here we have used *chain* to mean a path containing only forward arcs, the length of the chain being obtained by adding its arc lengths.

While this is a purely combinatorial problem, it may also be viewed as a flow problem simply by imposing a cost a_{ij} per unit flow in (i, j), taking all arc capacities infinite, and asking for a minimal cost flow of one unit from the first node to the second. An integral optimal flow corresponding to $v = 1$ singles out a shortest chain.

Many ways of locating shortest chains efficiently have been suggested. We describe one [10]. Like others, it simultaneously finds shortest chains from the first node to all others reachable by chains.

In this method each node i will initially be assigned a number π_i. These node numbers, which we shall refer to as *potentials*, will then be revised in an iterative fashion. Let 1 be the first node. To start, take $\pi_1 = 0$, $\pi_i = \infty$ for $i \neq 1$. Then search the list of arcs for an arc (i, j) whose end potentials satisfy

(1.1) $$\pi_i + a_{ij} < \pi_j.$$

(Here $\infty + a = \infty$.) If such an arc is found, change π_j to $\pi_j' = \pi_i + a_{ij}$, and search again for an arc satisfying (1.1), using the new node potentials. Stop the process when the node potentials satisfy

(1.2) $\pi_i + a_{ij} \geq \pi_j$

for all arcs.

It is not hard to show that the process terminates, and that when this happens, the potential π_j is the length of a shortest chain from 1 to j. (Here $\pi_j = \infty$ at termination means there is no chain from 1 to j.) A shortest chain from 1 to j can be found by tracing back from j to 1 along arcs satisfying (1.2) with equality (see Figure 1.1).

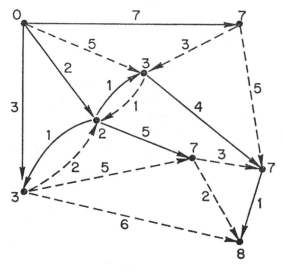

FIG. 1.1

Practical applications that require shortest chains are numerous. For instance, in making up a table of highway distances between cities, a shortest chain between each pair needs to be found. A less obvious application is the discrete version of the problem of determining the least time for an airplane to climb to a given altitude [2]. Some other applications will be discussed in following sections.

While we have assumed positive lengths for the method described above, this assumption can be weakened. Call a chain of arcs leading from a node to itself a *directed cycle*. Then it is enough to suppose that all directed cycle lengths are nonnegative.

If directed cycle costs are nonnegative, the minimum cost flow problem of Part I, section 3, can be solved by repeatedly finding cheapest chains in suitable networks. Because of the assumption on the cost function a, we may start with the zero flow. Thus, using Theorem 3.1, it is enough to reduce the problem of finding a cheapest flow augmenting path with respect to a minimal cost flow x of amount v to that of finding a cheapest chain. Define a new network $G' = [N; \alpha']$

from the given one $G = [N; \alpha]$ and the flow x as follows. First note that we may assume $x_{ij} \cdot x_{ji} = 0$, since $a_{ij} + a_{ji} \geqq 0$. Now put (i, j) in α' if either $x_{ij} < c_{ij}$ or $x_{ji} > 0$ and define a' by

$$(1.3) \qquad a'_{ij} = \begin{cases} a_{ij} & \text{if } x_{ij} < c_{ij} \text{ and } x_{ji} = 0, \\ -a_{ji} & \text{if } x_{ji} > 0. \end{cases}$$

Thus a chain from source to sink in the new network corresponds to an x-augmenting path in the old, and these have the same cost. Moreover, since x is a minimal cost flow, the function a' satisfies the nonnegative directed cycle condition. Hence the method of this section can be used to construct minimal cost flows of successively larger amounts.

2. Optimal chains in acyclic networks. If the network is acyclic (contains no directed cycles), the shortest chain method of the last section can be modified in such a way that, once a potential is assigned a node, it remains unchanged. One can begin by numbering the nodes so that if (i, j) is an arc, then $i < j$. Such a numbering can be obtained as follows. Since the network is acyclic, there are nodes having only outward-pointing arcs. Number these nodes $1, 2, \cdots, k$ in any order. Next delete these nodes and all their arcs, search the new network for nodes having only outward-pointing arcs, and number these, starting with $k+1$. Repetition of this process leads to the desired kind of numbering (see Figure 2.1).

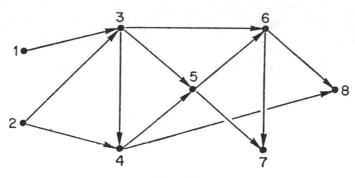

FIG. 2.1

If we wish to find shortest chains from node k to all other nodes reachable from k by chains, the calculation is now trivial. Simply define $\pi_k, \pi_{k+1}, \cdots, \pi_n$ recursively by

$$(2.1) \qquad \begin{cases} \pi_k = 0 \\ \cdots \\ \pi_j = \min_{k \leqq i < j} (\pi_i + a_{ij}), \quad j = k+1, \cdots, n. \end{cases}$$

Here the minimum is of course taken over i such that (i, j) is an arc.

Longest chains in acyclic networks can be computed by replacing "min" by "max" in (2.1).

The recursion (2.1), of dynamic programming type, can be applied in a number of problems. We shall discuss three such applications in the following sections.

3. The knapsack problem. Suppose there are K objects, the i-th object having weight w_i and value v_i, and that it is desired to find the most valuable subset of objects whose total weight does not exceed W. Thus we wish to maximize

$$(3.1) \qquad \sum_{i \text{ in } S} v_i$$

over subsets $S \subseteq \{1, 2, \cdots, K\}$ such that

$$(3.2) \qquad \sum_{i \text{ in } S} w_i \leq W.$$

We take $w_1, w_2, \cdots, w_K, W$ to be positive integers.

This combinatorial problem, commonly referred to as the knapsack problem, can be viewed as one of finding a longest chain in a suitable acyclic network. Let the network have nodes denoted by ordered pairs (i, w), $i = 0, 1, \cdots, K$, $w = 0, 1, \cdots, W$. The node (i, w) has two arcs leading into it, one from $(i-1, w)$, the other from $(i-1, w-w_i)$, provided these exist. (See Figure 3.1.) The length of the first arc is zero; the other has length v_i. In addition we put in a starting node and join it to all of the nodes $(0, w)$ by arcs of length zero. Then chains from the starting node to (i, w) correspond to subsets of the first i objects whose total weight is at most w, the length of the chain being the value of the subset.

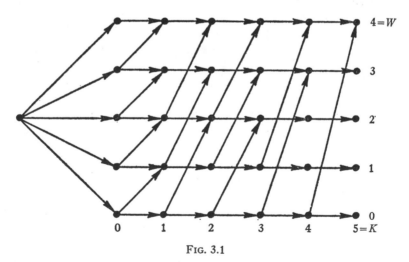

FIG. 3.1

4. Equipment replacement. As equipment deteriorates with age, and improved equipment becomes available on the market, a time may be reached when the purchase cost of new equipment is repaid by its potential future earnings. One is then faced with the problem of determining an optimal replacement policy [9].

For simplicity, consider a single machine and suppose that at the beginning of each of K periods of time it must be decided whether to keep the machine another period or purchase a new one. Let $r(i, t)$ denote the revenue obtainable during period i from a machine which starts the period at age t (the function r may reflect upkeep costs), and let $c(i, t)$ denote the cost of replacing a machine of age t with a new machine if the replacement occurs at the beginning of period i. Thus replacing a machine of age t at period i gives a net return for the period of $r(i, 0) - c(i, t)$.

The acyclic network shown in Figure 4.1 indicates one formulation in terms of chains. Again nodes are points (i, t), $i = 0, 1, \cdots, K, t = 0, 1, \cdots, T$. (Here T is some sufficiently large integer; if we start with a machine of age t, $T = K + t$ will do.) In general, two arcs, reflecting the possibilities of keeping or replacing, lead from (i, t), the "keep" arc going to $(i+1, t+1)$, the "replace" arc to $(i+1, 1)$. The first of these has length $r(i, t)$, the second has length $r(i, 0) - c(i, t)$. We may also put in a sink node with arcs of length zero leading into it from the nodes (K, t), $t = 0, 1, \cdots, T$. Then chains from $(0, t)$ to the sink correspond to the various replacement policies starting with a machine of age t, the length of the chain being the total return from the policy.

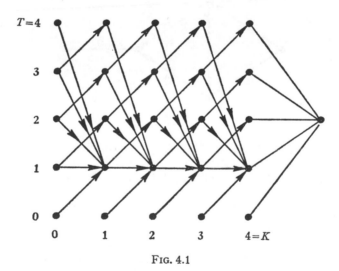

$T=4$

3

2

1

0

0   1   2   3   4=K

FIG. 4.1

A simpler network for this problem is shown in Figure 4.2. In Figure 4.2 an arc (i, j) corresponds to keeping the machine throughout periods $i, i+1, \cdots, j-1$ and replacing it at the start of period j, the associated length being the return obtained from this action. Thus a longest chain from source 1 to sink K is to be found.

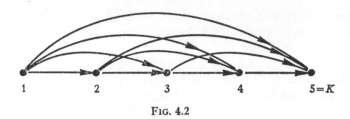

FIG. 4.2

The examples of this section and the preceding section are typical discrete dynamic programming problems. Such problems can always be viewed as seeking optimal chains in appropriate acyclic networks.

5. Project planning. One of the most popular combinatorial applications involving networks deals with the planning and scheduling of large, complicated projects [33]. Suppose that a project of some kind (the construction of a bridge, for example) is broken down into many individual jobs. Certain of these jobs will have to be finished before others can be started. We may depict the order relations among the jobs by means of an acyclic network whose arcs represent jobs. To take a simple case, suppose there are five jobs with the ordering: 1 precedes 3; 1 and 2 precede 4; 1, 2, 3 and 4 precede 5. This may be pictured by the network shown in Figure 5.1.

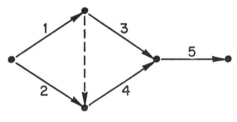

FIG. 5.1

Notice that we have added a "dummy" job, the dotted arc of Figure 5.1, to maintain the proper order relations among the jobs. The use of dummies permits a network representation of this kind for any project (finite partially ordered set).

Assuming that each job has a known duration time (dummies have zero duration time), that the only scheduling restriction is that all inward-pointing jobs at a node must be finished before any outward-pointing job can be started, it follows that the minimum time to complete the entire project is equal to the length of a longest chain of jobs. Hence the minimum project duration time can be calculated easily.

Although a fixed time has been assumed for each job, it may be the case that by spending more money, a job can be expedited. The question then arises: Which jobs should money be spent on and how much, in order that the project be finished by a given date at minimum cost? If the time-cost relation for each job is linear, this problem can be shown to be a minimal cost flow problem of the kind described in Part I, section 3 [18, 33].

6. **Minimal chain coverings of acyclic networks.** The following question concerning acyclic networks has both theoretical and practical interest: What is the minimum number of chains required to cover a given subset of arcs? We first show how flows may be used to answer this question, and then give a practical interpretation.

Let the subset of arcs be denoted by $\mathcal{Q}'$. To rephrase the question, we seek the minimum number of chains in the acyclic network G such that every arc of $\mathcal{Q}'$ belongs to at least one of these chains. Theorems 1.3 and 1.4 can be used to provide an answer to the question, as sources in G take all nodes with only outward-pointing arcs; as sinks take all nodes with only inward-pointing arcs. Now place a lower bound of 1 on flow in arcs of $\mathcal{Q}'$, 0 on arcs not in $\mathcal{Q}'$, and take all arc capacities infinite. Then an integral flow through G of amount v picks out v chains in G that cover all arcs of $\mathcal{Q}'$, and the second half of Theorem 1.4 implies

THEOREM 5.1. *The minimum number of chains in an acyclic network needed to cover a subset of arcs is equal to the maximum number of arcs of the subset having the property that no two belong to any chain.*

Theorem 5.1 is a generalization of a known result on chain decompositions of partially ordered sets [8].

A practical instance of this situation arises if we think of an airline, say, attempting to meet a fixed flight schedule with the minimum number of planes, all of the same type [4]. Let the individual flights be numbered 1, 2, $\cdots$, n. Start and finish times $s_i < f_i$ are known for each flight, and the times t_{ij} to return from the destination of the ith flight to the origin of the jth flight are also known. The flights can be partially ordered by saying that i precedes j if $f_i + t_{ij} \leq s_j$, and the resulting partially ordered set represented by an acyclic network (as in the preceding section). A chain in the network represents a possible assignment of flights to one aircraft. The problem then is to cover the nondummy arcs (those corresponding to actual flights) with the minimum number of chains. Theorem 5.1 asserts that this number is equal to the maximum number of flights, no two of which can be accomplished by a single plane.

Problems of this nature become considerably more complicated if the assumption of a fixed schedule is dropped. For instance, suppose the times s_i, f_i are at our disposal subject to the restriction that $f_i - s_i = t_i$, with the duration times t_i known, as well as the reassignment times t_{ij}. The problem might then be to arrange a schedule completing all flights by a given time and requiring the minimum number of planes, or to finish all flights at the earliest possible time with a fixed number of planes. For such scheduling problems there is very little known in the way of general theoretical results or good computational procedures. However, some special results have been deduced [27, 31].

7. Assignment problems. The following is typical of an important and well-known class of combinatorial problems having network flow formulations. Suppose there are m men and n jobs, and that it is known whether or not man i is qualified to fill job j, $i = 1, 2, \cdots, m, j = 1, 2, \cdots, n$. When is it possible to fill all jobs with qualified men and how does one determine such an assignment?

Using Theorem 1.3, the problem may be phrased in terms of flows. Corresponding to man i take a source node i, to job j a sink node j, and direct an arc from i to j if man i is qualified for job j. (See Figure 7.1.) Impose a demand of 1 unit at each sink and let each source have a supply of 1 unit. All arc capacities may be taken infinite. The problem of assigning men to jobs thus becomes that of constructing a flow (integral, of course) meeting the demands from the supplies.

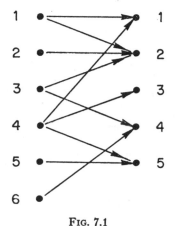

FIG. 7.1

Combinatorial interpretations of Theorems 1.1 and 2.1 for this situation lead in the first instance to a well-known theorem about maximum matchings and minimum covers in bipartite networks [35], and in the second instance to an equally well-known, and equivalent, theorem concerning systems of distinct representatives for subsets of a given set [24].

A more general assignment problem, usually referred to as that of optimal assignment [36, 42], assumes that man i in job j is worth a_{ij} units, and the total worth of an assignment is given by the sum of the numbers a_{ij} taken over the individual man-job matchings in the assignment. The problem then is to construct an assignment of maximal worth. By taking the cost per unit of flow in arc (i, j) to be $-a_{ij}$, the optimal assignment problem is seen to be a special case of the minimal cost flow problem.

More complicated personnel assignment models have been formulated in terms of flow networks. For instance, one which involves the recruiting, training, and retraining of military personnel to meet stipulated requirements in various job specialties over time has been treated in this way [23].

Applications of the optimal assignment model to other kinds of problems are very numerous. We mention one which involves the optimal depletion of in-ventory [7]. Suppose a stockpile consists of m items of the same kind, and that the age t_i of item i is known. Also known is a function $u(t)$ giving the utility for an item of age t when withdrawn from the stockpile, together with a schedule of demands specifying the times at which items will be required. The problem is to determine that order of item issue which maximizes the total utility while meeting the demand schedule. (For a concrete example, suppose one has m bottles of wine in his cellar, the ages of each being known, and consumes one bottle of wine weekly. The utility function for wine might appear as in Figure 7.2.) The utility of item i issued at time j is given by $u_{ij} = u(t_i + j)$, and hence the problem is to find an assignment of items to times which is optimal in terms of the u_{ij}.

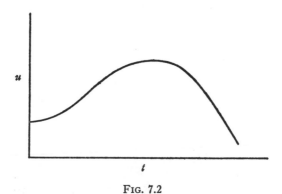

FIG. 7.2

If the utility function is convex, there is a simple rule for solving the prob-lem: Issue the youngest item first, then the next youngest, and so on. This policy, sometimes called LIFO (last in first out), may be shown optimal here by a simple interchange argument. Similarly, if the utility function is concave, the reverse rule FIFO (first in first out) solves the problem. In general, however, no such simple rule works and an optimal assignment needs to be computed.

8. **Production and inventory planning.** Problems involving dynamic produc-tion and inventory programs for a single type item have received considerable study. A very simple deterministic problem in this category is the following. Suppose there are n periods of time with known period demands $b_1, b_2, \cdots, b_n$ for the item, that the unit cost of production in period i is p_i, and the unit cost of storage from period i to $i+1$ is s_i. What pattern of production and storage meets the demands at minimum cost?

The network shown in Figure 8.1 assumes that production in period i can be used to satisfy demand in period i. The ith "production" arc (source arc) has infinite capacity and cost p_i; the ith "storage" arc has infinite capacity and cost s_i; the ith "demand" arc (sink arc) has capacity b_i and zero cost. The problem then is to determine a flow of amount $v = \sum_i b_i$ from source to sink that min-

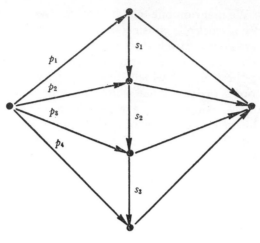

FIG. 8.1

imizes cost. Clearly production and storage capacities may be introduced if de-sired. But if these are left infinite, there is a very simple rule for solving the problem. For the ith demand, compare the chain costs

$$\begin{cases} p_1 + s_1 + \cdots + s_{i-1} \\ p_2 + s_2 + \cdots + s_{i-1} \\ \quad \cdots \\ p_i \end{cases}$$

and take the smallest of these. Then send b_i units of flow along the correspond-ing chain. An almost equally simple rule works in case each period's production cost is convex in the number of items produced [30].

If it is assumed that demands do not have to be satisfied, but that unful-filled demand in period i results in a penalty cost c_i per unit, we may place a flow cost $-c_i$ on the ith sink arc and solve the minimal cost flow problem parametri-cally in the flow amount v, selecting that v which gives the least cost.

9. Optimal capacity scheduling. The model of this section, proposed and studied in [41], is a rather general one which can be shown to include several of those previously discussed here. One version of the model is described in [41] as follows: "A decision maker must contract for warehousing capacity over n time periods, the minimal capacity requirement for each period being de-terministically specified. His economic problem arises because savings may pos-sibly accrue by his undertaking long-term leasing or contracting at favorable periods of time, even though such commitments may necessitate leaving some of the capacity idle during several periods."

To put the problem mathematically, let d_i be the minimal capacity demand in period i. Let x_{ij}, $i < j$, be the number of units of capacity acquired at the beginning of period i, available for possible use during periods i, $i+1$, $\cdots$, $j-1$, and relinquished at the beginning of period j, and let a_{ij} be the associated unit cost. Then the problem is to find $x_{ij} \geq 0$ that minimize

(9.1)
$$\sum_{i=1}^{n} \sum_{j=i+1}^{n+1} a_{ij} x_{ij}$$

subject to the constraints

(9.2)
$$\sum_{i=1}^{k} \sum_{j=k+1}^{n+1} x_{ij} \geq d_k, \qquad k = 1, 2, \cdots, n.$$

To see that the constraints (9.2) describe flows, first rewrite (9.2) as

(9.3)
$$\sum_{i=1}^{k} \sum_{j=k+1}^{n+1} x_{ij} - y_k = d_k, \qquad y_k \geq 0.$$

Next successively subtract the $(k-1)$-st equation from the kth, $k=n$, $n-1$, $\cdots$, 2, to obtain an equivalent system of constraints. The result is

$$\sum_{j=2}^{n} x_{ij} - y_1 = d_1,$$

$$\cdots$$

(9.4)
$$\sum_{j=k+1}^{n} x_{kj} - \sum_{i=1}^{k-1} x_{ik} + y_{k-1} - y_k = d_k - d_{k-1}, \qquad k = 2, \cdots, n,$$

$$\cdots$$

$$-\sum_{i=1}^{n-1} x_{in} + y_n = -d_n,$$

subject to which the linear form (9.1) is to be minimized.

The corresponding network is shown in Figure 9.1 below.

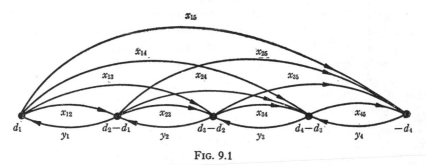

FIG. 9.1

Here x_{ij} is the flow in (i, j) and a_{ij} is the cost per unit of flow; y_i is the flow in $(i+1, i)$, with $a_{i+1,i}=0$. Nodes i for which $d_i - d_{i-1} > 0$ are sources with supplies $d_i - d_{i-1}$; nodes i for which $d_i - d_{i-1} < 0$ are sinks with demands $-(d_i - d_{i-1})$.

Referring to Figure 4.2, Part II, and Figure 9.1 above, it is apparent that the equipment replacement problem can be viewed as a special case of capacity scheduling by taking $d_i = 1$, all i.

A number of other situations that can be interpreted in terms of capacity scheduling are described in [41]. Mentioned are models involving checkout and replacement of stochastically failing mechanisms; determination of economic lot sizes, product assortment, and batch queuing policies; labor-force planning; and multi-commodity warehousing decisions.

One application (the dynamic economic lot size model [43]) deals with the problem described in section 8, where production costs now are concave functions of the number of items produced, and demands must be satisfied. Generally speaking, concavity makes minimization problems difficult, but here it can be seen that it is uneconomical to both produce in a period and carry inventory into the period, and hence there is an optimal policy of the following kind. The total time interval is broken into subintervals, with enough production at the beginning of each of these to satisfy its aggregate demand. Thus finding an optimal policy can be formulated in terms of capacity scheduling by letting a_{ij} be the total cost (including storage) associated with producing enough in period i to satisfy the demands for periods $i, i+1, \cdots, j-1$, and by taking all $d_i = 1$. In short, the problem has been reduced to one of finding a cheapest chain from source to sink in the network of Figure 9.1.

10. Minimal spanning trees. A network combinatorial problem for which there is a particularly simple solution method is that of selecting a minimum spanning subtree from an undirected network each of whose links has a length or cost. We may illustrate this problem with the following example. Imagine a number n of cities on a map and suppose that the cost of installing a communication link between cities i and j is $a_{ij} = a_{ji} \geq 0$. Each city must be connected, directly or indirectly, to all others, and this is to be done at minimum total cost. Clearly attention can be confined to trees (acyclic and connected networks of links), for if a connected network contains a cycle, removing one link of the cycle leaves the network connected and reduces cost. A minimal cost tree can be found easily as follows [37]. Select the cheapest link, then the next cheapest, and so on, being sure at each stage that no subset of the selected links forms a cycle. After $n-1$ selections, a cheapest tree has been constructed.

For example, in the network of Figure 10.1, this procedure might lead to the minimal cost tree shown in heavy links.

While it is not difficult to prove that this method solves the problem, it is nonetheless remarkable that being greedy at each stage works. There are few extremal combinatorial problems for which it does.

There is an interesting relation between the minimal spanning tree problem and another, which sounds on the surface to be very different. Think of the network in Figure 10.1 as being a highway map, where the number recorded beside each link is the maximum elevation encountered in traversing the link. Suppose someone who plans to drive from i to j dislikes high altitudes and hence

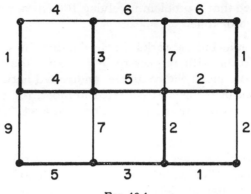

FIG. 10.1

wants to find a path connecting i and j that minimizes the maximum altitude. This problem is related to the shortest chain problem in the sense that methods for solving the latter are easily modified to solve the former, and this is so in either the directed or undirected case [40]. But it is also true in the undirected case that the minimal spanning tree solves the problem, and for all pairs of cities. That is, the unique path in the minimal spanning tree joining a pair of cities minimizes the path height [28]. Here we have used "path height" to mean the maximum number on the path.

There is also a min-max theorem concerning paths and cuts for this problem. Call the minimum link number in a cut the "cut height." Then it may be verified that the minimum height of paths joining two nodes is equal to the maximum height of cuts separating the two.

11. The traveling-salesman problem. Many problems that involve minimal connecting networks, and hence superficially resemble that of the last section, have no known simple solution procedures. For example, consider a network connecting a number of cities, with the length of each link being known, and imagine a traveling salesman who must start at some city, visit each of the others just once, and then return to the starting city. How does the salesman determine an itinerary that minimizes the total distance traveled?

A cycle that passes through every node of a network just once is usually called a Hamiltonian cycle. For brevity, we refer to it as a tour. Thus the problem asks for a shortest tour. (Of course the given network may contain no tour, but the existence question can be subsumed by considering all possible links present, those not corresponding to original ones having very large lengths.)

Not a great deal is known, either theoretically or computationally, about this problem, except that it is hard. On the theoretical side, for instance, there seem to be no simple conditions that are necessary and sufficient for a given network to contain a tour. On the computational side, while many methods for determining a shortest tour have been proposed, it is safe to assert that no one of these would guarantee that a problem involving 100 cities, say, could be solved in a reasonable length of time.

12. Minimal k-connected networks. In considering the synthesis of reliable communication networks with respect to link failure, the following question may be raised. Suppose given the complete, undirected network G on n nodes, where again each link of G has an associated number, the cost of installing a communication link between its end-nodes. For each $k = 1, 2, \cdots, n-1$, find a minimal cost k-link-connected spanning subnetwork of G [20]. Here a k-link-connected network is one in which at least k links must be suppressed in order to disconnect the network. Another way of characterizing this property is to say that every pair of nodes is joined by at least k link-disjoint paths. Thus k might be thought of as the "reliability level" of the communication network, and the practical problem is to minimize cost while achieving a stipulated reliability level.

For $k = 1$, the problem is that of section 10, and hence is readily solved. For $k = 2$ and all link costs 1 or ∞, the problem becomes that of determining whether a given network (the subnetwork of unit cost links) contains a tour, and is thus already difficult. But if all link costs are equal, the answer is known. Here the problem is to determine the minimum number of links required in a k-link-connected network on n nodes. For $k \geq 2$, there is an obvious lower bound on the number needed, namely $kn/2$ (for even kn) or $kn+1/2$ (for odd kn). These bounds can always be achieved.

This research was sponsored by the United States Air Force under Project RAND—Contract No. AF49 (638)-700.

References

1. R. G. Busacker and P. J. Gowen, A procedure for determining a family of minimal cost network flow patterns, O.R.O. Technical Paper, 15 (1961).

2. T. F. Cartaino and S. E. Dreyfus, Application of dynamic programming to the airplane minimum time-to-climb problem, Aero. Engr. Rev., 16 (1957) 74–77.

3. G. B. Dantzig, Application of the simplex method to a transportation problem, *Activity Analysis of Production and Allocation*, Cowles Commission Monograph 13, Wiley, New York, (1951) 359–373.

4. G. B. Dantzig and D. R. Fulkerson, Minimizing the number of tankers to meet a fixed schedule, Naval Res. Logist. Quart., 1 (1954) 217–222.

5. ———, On the max-flow min-cut theorem of networks, *Linear Inequalities and Related Systems*, Ann. of Math. Study 38, Princeton Univ. Press, (1956) 215–221.

6. G. B. Dantzig, D. R. Fulkerson, and S. Johnson, Solution of a large scale traveling salesman problem, Op. Res. 2, 393–410.

7. C. Derman and M. Klein, A note on the optimal depletion of inventory, Management Sci., 5 (1959) 210–214.

8. R. P. Dilworth, A decomposition theorem for partially ordered sets, Ann. of Math., 51 (1950) 161–166.

9. S. E. Dreyfus, A generalized equipment replacement study, J. Soc. Indust. Appl. Math., 8 (1960) 425–435.

10. L. R. Ford, Jr., Network flow theory, The RAND Corp., P-923 (1956).

11. L. R. Ford, Jr. and D. R. Fulkerson, Maximal flow through a network, Canad. J. Math., 8 (1956) 399–404.

12. ———, A simple algorithm for finding maximal network flows and an application to the Hitchcock problem, Canad. J. Math., 9 (1957) 210–218.

13. ———, A suggested computation for maximal multi-commodity network flows, Management Sci., 5 (1958) 97–101.

14. ———, Constructing maximal dynamic flows from static flows, Op. Res., 6 (1958) 419–433.

15. ———, *Flows in networks*, Princeton Univ. Press, 1962, 194 p.

16. D. R. Fulkerson, Increasing the capacity of a network: The parametric budget problem, Management Sci., 5 (1959) 472–483.

17. ———, An out-of-kilter method for minimal cost flow problems, J. Soc. Indust. Appl. Math., 9 (1961) 18–27.

18. ———, A network flow computation for project cost curves, Management Sci., 7 (1961) 167–178.

19. ———, A network flow feasibility theorem and combinatorial applications, Canad. J. Math., 11 (1959) 440–451.

20. D. R. Fulkerson and L. S. Shapley, Minimal k-arc-connected graphs, The RAND Corp., P-2371 (1961) 11 p.

21. D. Gale, A theorem on flows in networks, Pac. J. Math., 7 (1957) 1073–1082.

22. R. E. Gomory and T. C. Hu, Multi-terminal network flows, J. Soc. Indust. Appl. Math., 9 (1961) 551–571.

23. W. Gorham, An application of a network flow model to personnel planning, The RAND Corp., RM-2587 (1960) 85 p.

24. P. Hall, On representatives of subsets, J. Lond. Math. Soc., 10 (1935) 26–30.

25. F. L. Hitchcock, The distribution of a product from several sources to numerous localities, J. Math. and Phys., 20 (1941) 224–230.

26. A. J. Hoffman, Some recent applications of the theory of linear inequalities to extremal combinatorial analysis, Proc. Symposia Applied Math., 10 (1960).

27. T. C. Hu, Parallel sequencing and assembly line problems, Op. Res., 9 (1961) 841–849.

28. ———, The maximum capacity route problem, Op. Res., 9 (1961) 898–900.

29. W. S. Jewell, Optimal flow through networks with gains, Proc. Second International Conf. on Oper. Res., Aix-en-Provence, France 1960.

30. S. M. Johnson, Sequential production planning over time at minimum cost, Man. Sci., 3 (1957) 435–437.

31. ———, Optimal two- and three-stage production schedules with setup times included, Naval Res. Log. Q., 1 (1954) 61–68.

32. L. Kantorovitch and M. K. Gavurin, The application of mathematical methods in problems of freight flow analysis, *Collection of Problems Concerned with Increasing the Effectiveness of Transports*, Publ. of the Akad. Nauk SSSR, Moscow-Leningrad (1949) 110–138.

33. J. E. Kelley, Critical path planning and scheduling: mathematical basis, Op. Res., 9 (1961) 296–321.

34. T. C. Koopmans and S. Reiter, A model of transportation, *Activity Analysis of Production and Allocation*, Cowles Commission Monograph 13, Wiley, New York, 1951, 229–259.

35. D. König, *Theorie der endlichen und unendlichen Graphen*, Chelsea, 1950, 258 p.

36. H. W. Kuhn, The Hungarian method for the assignment problem, Naval Res. Log. Q., 2 (1955) 83–97.

37. J. B. Kruskal, Jr., On the shortest spanning subtree of a graph and the traveling salesman problem, Proc. Amer. Math. Soc., 7 (1956) 48–50.

38. G. J. Minty, Monotone networks, Proc. Roy. Soc., A, 257 (1960) 194–212.

39. A. Orden, The transshipment problem, Man. Sci., 3 (1956) 276–285.

40. M. Pollack, The maximum capacity route through a network, Op. Res., 8 (1960) 733–736.

41. A. F. Veinott, Jr. and H. M. Wagner, Optimal capacity scheduling—I and II, Op. Res., 10 (1962) 518–547.

42. D. F. Votaw, Jr. and A. Orden, The personnel assignment problem, Project SCOOP, Manual 10 (1952) 155–163.

43. H. M. Wagner and T. M. Whitin, Dynamic version of the economic lot-size model, Man. Sci., 5 (1958) 89–96.

8. AN APPRAISAL OF SOME SHORTEST-PATH ALGORITHMS

Stuart E. Dreyfus

University of California, Berkeley, California

(Received October 15, 1968)

This paper treats five discrete shortest-path problems: (1) determining the shortest path between two specified nodes of a network; (2) determining the shortest paths between all pairs of nodes of a network; (3) determining the second, third, etc., shortest path; (4) determining the fastest path through a network with travel times depending on the departure time; and (5) finding the shortest path between specified endpoints that passes through specified intermediate nodes. Existing good algorithms are identified while some others are modified to yield efficient procedures. Also, certain misrepresentations and errors in the literature are demonstrated.

IN THE never-ending search for good algorithms for various discrete shortest-path problems, some authors have apparently overlooked or failed to appreciate previous results. Consequently, certain recently reported procedures are inferior to older ones. Also, occasionally, inefficient algorithms are adapted to new, generalized problems where more appropriate modifiable methods already exist. Finally, the literature contains some erroneous procedures. This paper briefly considers various versions of discrete-path problems in the light of known results and some original ideas.

Our observations are, of course, by no means definitive or final. However, it is hoped that our somewhat skeptical survey of current literature will put the interested reader on guard and perhaps save him, or his digital computer, considerable time and trouble. Since our objective is more to alert than to resolve conclusively, this paper is informal and, at times, cryptic. We hope that even our most laconic remarks will prove enlightening for any reader deeply involved with the particular procedure.

Corresponding to almost any shortest-path algorithm, some special network structure exists for which the algorithm is efficient. Consequently, to give meaning to our conclusions, we shall generally restrict our attention to problems in which every pair of nodes is connected by an arc (perhaps of infinite length), and shall develop bounds on the number of computational steps. (For networks with special structures, the decomposition algorithms of references 1, 2, and 3 can prove effective.) One procedure is considered significantly superior to another when these bounds differ by a multiplicative factor involving N, the number of nodes. When the resulting formulas differ by only a multiplicative constant, the user's choice of algorithm should be determined by problem structure, computer configuration, and programming language.

1. THE SHORTEST PATH BETWEEN A SPECIFIED PAIR OF NODES

GIVEN A set of N nodes, numbered arbitrarily from 1 to N, and the $N \times N$ matrix D, not necessarily symmetric, whose element d_{ij} represents the length of the directed arc connecting node i to node j, find the path of shortest length connecting node 1 and node N. *Assume initially that* $d_{ii} = 0$ *and* $d_{ij} \geqq 0$. If no arc is directed from node i to node j, then $d_{ij} = \infty$; or, for purposes of digital computation, d_{ij} is taken large.

The computationally most efficient procedure was described first by DIJKSTRA[4] in 1959, and in 1960 by WHITING AND HILLIER (see reference 5, last paragraph beginning on page 39). The algorithm assigns tentative labels, which are upper bounds on the shortest distance from node 1, to all nodes; after the fundamental iterative step described below is repeated exactly once for each node, the tentative node labels are all permanent and represent shortest distances.

Initially, label node 1 with the permanent value zero, and tentatively label all other nodes infinity. Then, one by one, compare each node label except that at node 1 with the sum of the label of node 1 (i.e., 0) and the direct distance from node 1 to the node in question. The smaller of the two numbers is the new tentative label.

Next, determine the smallest of the $N-1$ tentative labels and declare it permanent. Suppose that node k is the one permanently labeled. Then, one at a time, compare each of the $N-2$ remaining tentative node labels to the sum of (a) the label just assigned permanently to node k and (b) the direct distance from node k to the node under consideration. The smaller of the two numbers becomes the tentative label. Determine the minimum of the $N-2$ tentative labels, declare it permanent, and make it the basis of another modification of the remaining tentative labels of the type described above. When, after at most $N-1$ executions of the fundamental iterative step, node N is permanently labeled, the procedure terminates. (If the shortest paths from node 1 to all other nodes are desired, the fundamental iterative step must be executed exactly $N-1$ times.)

The optimal paths can easily be reconstructed if an optimal policy table (in this case, a table indicating the node from which each permanently labeled node was labeled) is recorded. Alternatively, no policy table need be constructed, since it can always be determined from the final node labels by ascertaining which nodes have labels that differ by exactly the length of the connecting arc (*see* Note 1).

The proof of the validity of the method is inductive, with the key step as follows. Suppose that, at a particular stage, the nodes are divided into two mutually exclusive and collectively exhaustive sets—Set 1 contains the permanently labeled nodes and Set 2 the temporarily labeled ones. The node labels of Set 1 are correct minimum distances from the source node. The node labels of Set 2 are the shortest distances from the source that can be attained by a path in which all except the terminal node belong to Set 1. Then the minimum-label node of Set 2—call it node i—can be transferred to Set 1, because if a shorter path to node i existed it would have to contain a first node that is currently in Set 2. However, that node must be farther away from the source, since its label exceeds that of node i; and the continuation path to node i must have nonnegative length since we have assumed that all distances are nonnegative. The subsequent use of node i to reduce the labels of adjacent nodes belonging to Set 2 restores to Set 2 the property assumed above.

This algorithm requires $N(N-1)/2$ additions and $N(N-1)$ comparisons to solve the problem—totaled over all steps, $N(N-1)/2$ additions and comparisons are necessary to compute tentative node labels and $N(N-1)/2$ comparisons are necessary to find the minimum label at each step. All steps are naturally and easily programmed except that of distinguishing which nodes are permanently and which tentatively labeled. Some computational experimentation indicates that an efficient method of distinguishing is to attach to each node an index number that changes from, say, 0 to 1 when a node label becomes permanent. When branching out from a just-permanently-labeled node, the index of the destination node is consulted as each outgoing arc is considered. If the index is zero, the temporary label of the destination is reduced, if appropriate. At the same time, memory cells designating the smallest temporary label encountered thus far during the branching and the associated destination node are modified, if appropriate. This programming device requires $(N-1)^2$ comparisons to consult the indices. Hence, a total of about $N^2/2$ additions and $2N^2$ comparisons is necessary.

POLLACK AND WIEBENSON (*see* reference 6, p. 225) describe and credit to MINTY (*see* Note 2) the first (and hence of some historical interest) systematic, easily programmed, permanent-label-setting precursor of the above method. That algorithm is probably due originally to FORD AND FULKERSON,[7] who developed it for a more general flow problem, of which the shortest-path problem is a special case requiring the fastest flow of one item. The method requires approximately $N^3/6$ additions and comparisons for solving the shortest-path problem, and hence is not recommended.

The Minty-Ford-Fulkerson procedure can be accelerated—but not enough to compete in general with the Dijkstra algorithm—by using a modification reported by Whiting and Hillier (the first method of reference 5) and DANTZIG.[8] In their method, derived independently of each other and of the Minty and Ford-Fulkerson[7] works, outgoing arcs from each node are listed from shortest to longest. This obviates a search for the shortest arc out of each permanently labeled node at each step, and reduces the number of additions and comparisons from $N^3/6$ to $N^2/2$. While this number suggests that the method improves the Dijkstra procedure, it is misleading for two reasons. First, ordering the data as required by the method necessitates approximately $N^2 \log_2 N$ additional comparisons. Second, the method must delete arcs from lists as they are used; and also, each time a node is permanently labeled, the method must delete, from all lists, arcs that lead into that node. This data modification requires

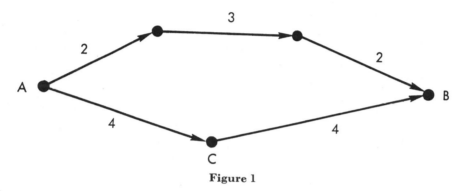

Figure 1

about $3N^2$ comparisons, plus elaborate programming of list-processing procedures. Therefore, except perhaps for sparse networks with far fewer than N^2 arcs, the method is not recommended.

Some authors have proposed simultaneously fanning out from both endpoints as a means of reducing computation. BERGE AND GHOUILA-HOURI[9] falsely assert that when, for the first time, some node is permanently labeled in both fans, the optimal path results and goes through that node. Dantzig[10] is vague about his stopping procedure. In the problem depicted in Fig. 1, if the Dijkstra scheme is first used to permanently label the node nearest A reachable from A, then the node nearest B from which B can be reached, then the second closest to A, etc., the node C is permanently labeled both 'out from A' and 'into B' after two applications of the procedure at each end. Yet ACB is *not* the shortest path.

When the correct stopping procedure given cryptically by NICHOLSON[11] and explained clearly by MURCHLAND[12] is used, the least upper bound on the computation required by a two-ended procedure exceeds that for the one-ended Dijkstra algorithm, contrary to assertions by BERGE AND GHOUILA-HOURI.[9] This happens because, as more nodes become permanently labeled in the Dijkstra procedure, a decreasing number of additions and comparisons is necessary to modify all tentative labels. As a result, determining the first $N/2$ permanent labels requires just over three-fifths the work of the complete solution. If the Nicholson stopping condition is satisfied long before $N/2$ nodes have been permanently labeled from each terminus, a saving may accrue; but, in a case where nearly all N nodes must be permanently labeled from either one end or the other, the two-ended procedure will prove inefficient.

All these methods require all elements of D to be nonnegative. A related problem assumes that some d_{ij} are negative, but the sum of the d_{ij} around any loop is positive. Such data arise when arc numbers represent costs, and some arcs are profitable. (The problem has no solution if negative loops exist. Should negative loops exist but be excluded from admissible paths, no known algorithm is satisfactory.)

We present, first, a well-known algorithm that either solves the problem, with negative d_{ij}, in at most N^3 additions and comparisons, or detects the existence of a negative cycle in that number of steps; second, a recent improvement that quarters the number of calculations; and third, a different procedure that is competitive. While we know of no better procedures, the disparity in methods and bounds indicates the probability of further improvements.

The basic procedure has been proposed, originally for problems with $d_{ij} \geq 0$, by Ford,[13] MOORE,[14] BELLMAN,[15] and undoubtedly others. This procedure repeatedly updates all node labels. For the initial condition given in equation (1), the node label $f_i^{(k)}$ represents the length of the shortest path that connects node 1 and node i and that contains $k+1$ or fewer arcs. Unlike the permanent-label-setting procedures recommended when all d_{ij} are nonnegative, in this procedure no node labels are considered final until all are. The fundamental recursion is

$$f_i^{(k+1)} = \min_j [d_{ji} + f_j^{(k)}], \qquad f_i^{(0)} = d_{1i}. \tag{1}$$

For the case $d_{ij} \geq 0$, convergence occurs whenever $f_i^{(k)} = f_i^{(k+1)}$ for all i, or after $N-2$ iterations if the former situation occurs no sooner (since no shortest path contains more than $N-1$ arcs). If $(N-2)$ iterations are required and if each pair of nodes is connected by an arc, then $(N-2)(N-1)^2$ additions and comparisons take place.

This method is inefficient for a positive-distance problem if two or more iterations of (1) are needed; and, unless all $N-2$ iterations are required, as many iterations will be necessary as the number of arcs in the shortest path from node 1 to node j, where node j is the node whose shortest path has the greatest number of arcs.

When this procedure is applied to a problem with some negative d_{ij}, either convergence will occur for $k \leq N-1$, indicating no negative cycles exist and the solution is optimal; or a change in some f_i will occur on the $(N-1)$st iteration, indicating the existence of a negative loop.

YEN[16] has recently reduced the computation by a factor of two. He suggests the recursion, for $k=1, 2, \cdots$,

$$f_i^{(2k-1)} = \min[\min_{j<i} (d_{ji}+f_j^{(2k-1)}), f_i^{(2k-2)}], \qquad (i=1, \cdots, N)$$
$$f_i^{(2k)} = \min[\min_{j>i} (d_{ji}+f_j^{(2k)}), f_i^{(2k-1)}], \qquad (i=N, \cdots, 1) \tag{2}$$

with initial condition $f_i^{(0)}=d_{1i}$. Using new values of f as soon as they are determined, and processing the nodes alternately forwards and backwards, produces bounds on convergence the same as above; yet, minimizing only over nodes previously treated necessitates only half as many additions and comparisons per iteration [note that in (1) the minimization is over all j while in (2) it is only over either $j<i$ or $j>i$]. In another paper, not yet published, Yen has shown how the computation can be reduced even further to $N^3/4$ by exploiting the fact that, at each iteration, one additional f_i becomes the correct minimum distance, and therefore affects the calculations no more thereafter.

A scheme of DANTZIG, BLATTNER, AND RAO[17] is novel and, with a slight modification, efficient. It can detect a negative loop much more quickly than the above procedures, if such a loop exists; it is as fast, if no such loop exists. Unfortunately, the procedure is difficult to explain, prove, or program.

Suppose that, at iteration $k+1$, node numbers $f_i^{(k)}$ have been assigned to nodes 1 through k, each representing the length of the shortest path to node i that may pass through intermediate nodes 1 through k, but no others. Also suppose that there are no negative cycles composed only of nodes 1 through k. The length of the shortest path to node $k+1$, using only nodes 1 through $k+1$, and not containing a negative cycle, is found by

$$f_{k+1}^{(k+1)} = \min_j [f_j^{(k)}+d_{j,k+1}]. \qquad (j=1, \cdots, k) \tag{3}$$

Then the node numbers $f_j^{(k)}$, $j=1, \cdots, k$, are reduced if, by introducing node $k+1$ into a path, a shorter distance results. This calculation involves a subiteration that first finds the node whose distance can be most reduced by introducing node $k+1$ into the path, then the node with the second-largest reduction, etc. This reduction is accomplished in a manner analogous to the Dijkstra procedure[4] except that an additional calculation is needed to see if the introduction of node $k+1$ has created a negative cycle (reference 17 gives a less efficient method than reference 4, the former involving list processing and reordering). At iteration $k+1$, (3) involves a negligible k additions and comparisions. The reduction of other node numbers based on node $k+1$, if carried out as in reference 4, requires at most $2k^2$ comparisons and k^2 additions. The latter number is double that ascribed to the Dijkstra algorithm in Sec. 1 because the computation of a potential node reduction involves a subtraction as well as an addition. The test for a negative cycle involves a negligible k additions and comparisons (the subiteration terminates when no further reduction is found, which may be well before the bounding number of calculations is performed). Hence for N nodes, the method appears to involve at most $N^3/3$ additions and $2N^3/3$ comparisons—computation comparable to the upper bound for the Yen[16] algorithm. Two totally different procedures that produce upper bounds on computational steps that are generally not realized and that differ only by a factor of 2 must be considered competitive.

In concluding our treatment of the problem of finding the shortest path between a specified source node and all other nodes of a network, let us briefly summarize the results of some computational experiments of HITCHNER[18] involving various specially structured problems, all with the restriction that $d_{ij} \geq 0$. Hitchner compares one method that successively reduces node labels by considering neighboring nodes [in the spirit of the algorithm described by (1)] with four variations on the permanent-label-setting Dijkstra procedure. Three of the latter keep lists to avoid repeated minimization over the set of all nonpermanently labeled nodes. He concludes that, for problems with only four arcs emanating from each node (such as in 'ideal' city maps), the procedure based on (1) and one special list-keeping version of the Dijkstra method are superior (because they depend more on the number of arcs than nodes). For problems with 25 per cent or more of the $N(N-1)$ possible arcs present, the Dijkstra procedure (with no sophisticated list-processing adornments) out-performed all competitors. (Reference 19 gives descriptions and details of dozens of minor variations on the two basically different procedures we have distinguished above, but no attempt at empirical comparison is made.)

2. THE SHORTEST PATHS BETWEEN ALL PAIRS OF NODES OF A NETWORK

Two SOMEWHAT different, but equally elegant and efficient, algorithms are recommended. One was published without comment as an obscure nine-line ALGOL algorithm in 1962 by FLOYD,[20] based on a procedure by WARSHALL,[21] and was rediscovered and appropriately extolled in 1965 by MURCHLAND.[22] The other was produced in 1966 by Dantzig.[23] Since both require *exactly* the same number of calculations—$N(N-1)(N-2)$

additions and comparisons for the case $d_{ij} \geq 0$—are easily proved, and programmed, and culminate a steady progression of successive improvements
(*see* references 24, 15, 25, and 26; actually reference 20 precedes the inferior algorithm of reference 26), there is good reason to believe that they
are definitive. While the reader should consult the primary sources cited
above, we briefly describe the algorithms here.

The Floyd procedure[20] builds optimal paths by inserting nodes, when
appropriate, into more direct paths. Starting with the $N \times N$ matrix D
of direct distances, N matrices are constructed sequentially. The kth
such matrix can be interpreted as giving the lengths of the shortest allow-
able paths between all node pairs (i, j), where only paths with intermediate
nodes belonging to the set of nodes 1 through k are allowed. The $(k+1)$st
matrix is constructed from the kth by using the formula

$$d_{ij}^{(k+1)} = \min[d_{ij}^{(k)}, d_{i,k+1}^{(k)} + d_{k+1,j}^{(k)}], \qquad d_{ij}^{(0)} = d_{ij}. \tag{4}$$

Here, k, which is initially zero, is incremented by 1 after i and j have ranged
over the values $1, \cdots, N$; and $k = N-1$ at termination.

To appreciate the rationale of the procedure, suppose the shortest path
from node 8 to node 5 is 8–3–7–1–9–5. Iteration 1 will replace d_{79} by
$d_{71} + d_{19}$; iteration 3 will replace the current value of d_{87} (which may or may
not be the original value) by $d_{83} + d_{37}$ (the optimal value); iteration 7 will
replace the current d_{89} by $d_{87} + d_{79}$, where these numbers are the optimal
values as computed above; and iteration 9 will obtain for d_{85} the sum of
d_{89} and d_{95}, when d_{89} is as computed at iteration 7. Hence, the correct
shortest distance is obtained. (The above justification is, of course, not a
proof.)

A minor modification, in case some d_{ij} are negative, detects negative
loops, if any exist, and otherwise yields correct results. An additional
advantage of this procedure is that $N-1$ additions and comparisons are
easily circumvented whenever an element $d_{i,k+1}^{(k)}$ equals infinity in (4), i.e.,
no path, with only nodes 1 through k as intermediate nodes, connects nodes
i and $k+1$ (*see* Note 3). As in the case of the particular initial-terminal
pair problem, an optimal policy table (matrix) associating with the initial-
terminal node pair (i, j) the next node along the best path from i to j
can be developed during the computation, or can be deduced from the final
shortest-distance matrix.

Dantzig's scheme[23] generates successive matrices of increasing size.
The kth iteration produces a $k \times k$ matrix whose elements are the lengths of
the shortest paths connecting nodes i and j, $i = 1, \cdots, k$, $j = 1, \cdots, k$, in
which only nodes 1 through k may be intermediate nodes. Given the
$k \times k$ matrix $D^{(k)}$ with elements $d_{ij}^{(k)}$ as defined above, compute $D^{(k+1)}$ as
follows:

(1) Compute $d_{i,k+1}^{(k+1)}$ for $i=1, \cdots, k$ by

$$d_{i,k+1}^{(k+1)} = \min_{1 \le j \le k} [d_{ij}^{(k)} + d_{j,k+1}],$$

and similarly compute $d_{k+1,i}^{(k+1)}$.

(2) Compute $d_{ij}^{(k+1)}$ for $i=1, \cdots, k, j=1, \cdots, k$ by

$$d_{ij}^{(k+1)} = \min[d_{ij}^{(k)}, d_{i,k+1}^{(k+1)} + d_{k+1,j}^{(k+1)}].$$

That the $d_{ii}^{(k)}$ yielded by these steps are as previously defined is obvious after a little thought. (Reference 23 gives a proof.) If all distances are positive, $d_{ii}^{(k)} = 0$ for all i and k. If not, $d_{ii}^{(k)}$ can be computed easily; and if any $d_{ii}^{(k)}$ is negative, a negative loop exists.

The Dantzig algorithm seemingly cannot exploit nonexistent arcs in a manner similar to Floyd's.

If the above algorithms, requiring $N(N-1)(N-2)$ additions and comparisons, are indeed as efficient as possible, then the most efficient particular-pair algorithm must require at least $(N-1)(N-2)$ such calculations (assuming, as was the case in Sec. 1, that such procedures must—at least in the worst case—generate best paths from the initial node to all other nodes). The best algorithm of Sec. 1, that which required no elaborate data-preparation or list keeping, involves $N^2/2$ additions and $2N^2$ comparisons. Assuming additions and comparisons use equal amounts of computation time, theory gives, as a reasonable upper-bound estimate of potential future improvements for the particular-pair-of-endpoints problem, a reduction in computation time of about 20 per cent. Actual computational experiments indicate that computing optimal paths between all pairs of nodes by N applications of the Dijkstra method requires roughly one and one-half times the time consumed by either algorithm specifically solving the all-pairs problem. Indexing operations account for the difference between theory and practice. Viewed from the perspective of a combinatorialist, the well is nearly dry. (Such may not be the case for the two procedures recommended in Sec. 1 for the problem with negative distances.)

References 2 and 3 specifically concern the exploitation, within the algorithms described in this section, of special structure.

3. DETERMINATION OF THE SECOND-SHORTEST PATH

IT IS occasionally desirable to know the value of the second- (and third-, etc.) shortest path through a network. For example, suppose that some complex quantitative (or even qualitative) feature characterizes paths, and the shortest path possessing this additional attribute is sought. By ignoring the special aspect in question and ordering paths from shortest to longest, the best path with the additional feature can sometimes be determined efficiently.

The analysis below initially considers the problem of determining the second-best path between a specified initial node 1 and a specified destination, N. Then it draws conclusions for more general problems. Two paths that do not visit precisely the same nodes in the same order are considered different. The discussion ignores ties by assuming, for simplicity, that all paths have different values. A path with a loop is considered an admissible path, and indeed such a path can be second best, even for problems with all $d_{ij}>0$. Even node N may be visited twice along the second-best path.

The earliest good algorithm known to this author was proposed by HOFFMAN AND PAVLEY.[27] A *deviation* from the shortest path was defined to be a path that coincides with the shortest path from its origin up to some node j on the path (j may be the origin or the terminal node), then deviates directly to some node k not the next node of the shortest path, and finally proceeds from k to the fixed terminal node via the shortest path from k. Reference 27 shows the second-shortest path between specified initial and terminal nodes to be a deviation from the shortest path.

To solve the problem posed above, first the shortest paths from all initial nodes to the specified destination are determined by means of any efficient algorithm. Then, all deviations from the shortest path between the specified origin and terminus are determined, evaluated, and compared, and the best noted. If the average node has M outgoing links, and the average shortest path contains K arcs, an average problem is solved in approximately MK additions and comparisons beyond those required for solution of the shortest-path problem.

Suppose second-shortest paths from all nodes to the specified terminal node N are sought. Then we propose the following modification of the Hoffman-Pavley method.[27] After solving the shortest-path problem, determine v_N, the length of the second-shortest path from N to N (it may be infinity), by considering all deviations at N. Then, for each node k whose shortest path to N contains only one arc, compare (a) the length of the shortest path deviating at k and (b) $d_{kN}+v_N$. The minimum of these two quantities is v_k, the length of the second-shortest path from that node. Then, consider all nodes j that are two arcs from N via the shortest path. For each, compare (a) the length of the shortest path deviating at j and (b) the length of the arc d_{ji} that is the first arc of the shortest path from j to N plus v_i, the previously-determined length of the second-shortest path from i to N. The minimum value is v_j. Repeat this iterative process until all nodes are labeled. Note that the iteration is performed on an index representing the number of arcs in the shortest path from each node. This procedure requires about MN additions and comparisons.

Reference 28, published subsequently to the above method, gives a seemingly different procedure. Define u_i as the length of the shortest path from node i to a specified terminal node N, and v_i as the length of the second-shortest path. Define $\min_k (x_1, \cdots, x_n)$ as the kth-smallest value of the quantities x_i. Then, according to reference 28, v_i is characterized by the equation

$$v_i = \min \left[\begin{array}{l} \min_2(d_{ij}+u_j) \\ {}_{j \neq i} \\ \min_1(d_{ij}+v_j) \\ {}_{j \neq i} \end{array} \right], \quad (i=1, \cdots, N-1) \ (5)$$

$$v_N = \min_i [d_{Ni}+u_i].$$

The term $\min_2 (d_{ij}+u_j)$ determines the value of the best path originating at node i and deviating from the shortest path at that node i. The term $\min_1 (d_{ij}+v_j)$ evaluates the best path consisting of any first arc, plus the second-best continuation. The originators of the method apparently did not notice that if the minimizing node in the $\min_1$ operation is not k (the next node of the already-known shortest path from i) but some other node p, then $d_{ip}+u_p$ (an admissible solution to the $\min_2$ expression) is less than $d_{ip}+v_p$ (since u_p v_p). Since the term $\min_1 (d_{ij}+v_j)$ can yield the over-all minimum in (5) only if $j=k$, where k is the node after i on the shortest path from i, the $\min_1$ term in (5) can be replaced by merely $d_{ik}+v_k$.

After this reduction by a factor of two in required computation, we can calculate the method's approximate computational requirements. BELL-MAN AND KALABA[28] recommend solution of (5) by an iterative procedure, where v_i is superscripted on the left by $(k+1)$ and v_j on the right by (k). Defining M and N as above and L as the average number of iterations until convergence of the iterative solution of (5), the method requires an average of NML additions and comparisons. L is less than $N-1$ and may be as large as the number of arcs in the shortest path containing the most arcs. However, after replacing the $\min_1$ term in (5) by $d_{ik}+v_k$ as discussed above, solution can be greatly accelerated by using a one-pass scheme first labeling nodes one-arc-by-shortest-path from N, then two-arcs-by-shortest-path, etc. This reduces the Bellman-Kalaba[28] procedure to precisely the modified Hoffman-Pavley[27] algorithm recommended above.

In summary, if only the second-shortest path connecting a *particular* pair of nodes is desired, the method of reference 27 is clearly best since it requires MK calculations compared to MN for our improved version of the method of reference 28, which must solve the all-initial-nodes problem in order to resolve the particular-initial-node case. If a problem involving a fixed terminal node and all possible initial nodes is posed, the methods as modified in this paper are equivalent. These conclusions contradict those of POLLACK's survey paper.[29]

To determine the third-shortest paths from all initial nodes to N, we recommend the following generalization of the above procedure. If w_i represents the length of the third-shortest path from i, then

$$w_i = \min \begin{bmatrix} d_{ik} + w_k \\ \min_2[d_{ij} + u_j] \\ {\scriptstyle j \neq i} \end{bmatrix}, \qquad (6)$$

if a single node k follows i along both the first- and second-shortest paths. If k is the node following i on the shortest path, and m is the node following i on the second-shortest, then

$$w_i = \min \begin{bmatrix} d_{ik} + v_k \\ d_{im} + v_m \\ \min_3[d_{ij} + u_j] \\ {\scriptstyle j \neq i} \end{bmatrix}. \qquad (7)$$

Once the functions u_i and v_i have been determined, the function w_i can be computed node by node by first computing w at nodes that are one arc distant from N via the shortest path, then two arcs, etc.

The general procedure can be stated as follows. Define $n_i(j, k)$ as the number of paths, among the k shortest from i to N, that begin by going from i to j. Define f_i^k as the length of the kth shortest path from i to N. Then

$$f_i^k = \min_{j \neq i} [d_{ij} + f_j^{n_i(j,k-1)+1}], \qquad (k = 2, 3, \cdots) \quad (8)$$

where, for the minimizing j, $n_i(j, k) = n_i(j, k-1) + 1$ and for all other j, $n_i(j, k) = n_i(j, k-1)$. Initially, $n_i(j, 1)$ equals 1 or 0, depending on the shortest path from i to N. As noted, except for the case $k = 1$, if the nodes are processed in order of increasing number of arcs in the shortest path to N, the right-hand side of (8) will always be known when the left-hand side is evaluated.

While we assumed above, in order to avoid complicating the explanations, that no two paths have the same length, all of these methods can be generalized—at the expense of a little additional bookkeeping—to include the treatment of possible ties. The problem of ties can sometimes be avoided by slightly perturbing the data.

Reference 30 examines an entirely different procedure, the efficiency of which is difficult to determine.

CLARKE, KRIKORIAN, AND RAUSEN[31] treat this problem under the additional restriction that only loopless paths are admissible. Their branch-and-bound algorithm involves generating, listing, and processing all paths with certain properties. The number of such paths is not easy to bound.

Certainly much of the elegance of the above algorithms is lost. It is unclear how the procedure of reference 31 generally compares with that of merely producing next-best paths (possibly containing loops) by the above algorithms until obtaining the desired number of loopless ones.

It is not simple to modify (5) correctly so as to exclude paths with loops (reference 29, page 555, explains why). While Pollack[29] clearly recognizes the problem, his subsequent scheme encounters circularity, since determination of the kth-best loopless path may depend on the length of certain other $k+p$, $p \geq 1$, best paths, and conversely. Another scheme of Pollack[32] is clearly correct, but computation increases rapidly with k and the method can be recommended for only very small k.

4. TIME-DEPENDENT LENGTHS OF ARCS

AT LEAST ONE paper[33] has studied the problem of finding the fastest path between cities where the time of travel between city i and city j depends on the time of departure from city i. When t is the time of departure from city i for city j, let $d_{ij}(t)$ denote the travel time. (If travel schedules are such that a delay before departure decreases the time of arrival, $d_{ij}(t)$ represents the elapsed time between time t and the earliest possible time of arrival.) This model has applications in the areas of transportation planning and communication routing.

COOKE AND HALSEY[33] define $f_i(t)$ as the minimum time of travel to N, starting at city i at time t, and establish the formula

$$f_i(t) = \min_{j \neq i} \{d_{ij}(t) + f_j[t + d_{ij}(t)]\}, \qquad f_N(t) = 0. \tag{9}$$

Assuming all the $d_{ij}(t)$ are defined at, and take on, positive integer values, an iterative procedure is given for finding the quickest paths from all cities to city N, starting at city i at time 0. Defining T to be the maximum taken over all i of $d_{iN}(0)$ (a smaller T can be determined, at some inconvenience, if this number is infinite), and assuming all cities are connected at all times (perhaps by links taking infinite time), the procedure requires at most $N^2 T^2$ additions and comparisons.

This problem can be solved by the method of Dijkstra[4] discussed in Sec. 1 just as efficiently as can the problem where the times (or distances) are not time-dependent. Also, the restriction to integer-valued times can be dropped and any real-valued times can be treated. Define the tentative node (city) label f_i to be an upper bound on the earliest time of arrival at node i, and permanent labels to be earliest possible (optimal) times-of-arrival. First, permanently label node i_0 (the initial node) zero and all other nodes infinity. Next, tentatively label all nodes j with the minimum of the current node label f_j and the sum of f_{i_0} and $d_{i_0 j}(f_{i_0})$. Then, find the minimum, nonpermanent node label, say f_k, and declare it permanent. (f_k

is the earliest possible time of arrival at node k, leaving node i_0 at time 0.) Node k is then used to try to reduce the labels at all tentatively labeled cities, by comparing $f_k + d_{kj}(f_k)$ to the current label, and the minimum new temporary label is made permanent, etc. After at most N^2 comparisons and $N^2/2$ additions, city N is labeled and, leaving i_0 at time 0, the quickest paths to all nodes, including N, are determined. As is the case for the closely related Dijkstra procedure,[4] about $N^2/2$ additions and $2N^2$ comparisons are required when implementing a method of distinguishing temporarily from permanently labeled nodes. If quickest paths from all cities to N are desired, the algorithm must be repeated $N-1$ times; but, even then, the procedure compares favorably, in both computation and required assumptions, with the Cooke-Halsey algorithm.[33]

5. SHORTEST PATHS VISITING SPECIFIED NODES

GIVEN A SET of N nodes and distances $d_{ij} \geq 0$, suppose we desire to find the shortest path between nodes 1 and N that passes through the $k-1$ nodes 2, 3, $\cdots$, $k \leq N-1$, called 'specified nodes.' A simple, but completely erroneous, solution of this problem was reported by SAKSENA AND KUMAR.[34] Noting this, we wish to give a solution method.

The fallacy in reference 34 is the assertion (subject to a proviso to follow) that the shortest path from a specified node i to N passing through at least p of the specified nodes enroute is composed of the shortest path from i to some specified node j, followed by the shortest path from j to N passing through at least $p-r-1$ specified nodes, where r is the number of specified nodes that lie on the shortest unrestricted path from i to j. Saksena and Kumar[34] incorrectly assert that this is true, provided—should specified nodes occurring on the shortest path from i to j also lie on the continuation path from j to N and therefore be counted twice—that at least p distinct specified nodes lie on the path. Should some candidate path corresponding to some j violate the duplication-of-nodes proviso above, that possibility is inadmissible and the possibility of going initially from i to j by shortest path is dropped from consideration. The procedure fails to note that, in this case, some less short continuation from j passing through at least $p-r-1$ specified nodes, and avoiding duplication of nodes, may yield a better path than the best remaining path satisfying the conditions described above. For example, in the network shown in Fig. 2, with all nodes considered specified, suppose we seek the best path from 1 to 5 passing through at least two intermediate nodes. The best path from 1 to 4 has length 3 and happens to pass through no nodes enroute, and the best continuation from 4 to 5 passing through at least one node has length 4; hence, this possibility has length 7. The best path from 1 to 3 has length 2 and happens to pass through no nodes enroute, and the best continuation from 3 to 5 passing through one node has length infinity (no such continuation from 3 exists). The best path from 1 to 2 has length 1 and happens to pass through no nodes enroute, and the best continuation from 2 to 5 passing through one node enroute has length 2 (it returns to node 1), yielding a sum of 3. However, it is inadmissible as a path through two intermediate specified nodes because node 1 is counted twice. The answer, by the Saksena-Kumar algorithm,[34] would then be 7, the best of the other alternatives. Yet, the path 1–2–3–5 has length 4 and is admissible. This is an example

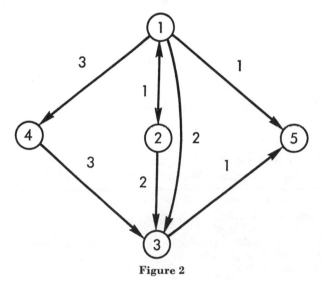

Figure 2

of a best first portion and a second-best continuation being optimal. (Or, the same path can alternatively be viewed as the second-best path from 1 to 3 followed by the best continuation.) No simple modification of the referenced method seems to handle this kind of situation.

Assuming paths with loops are admissible, the problem can be correctly solved as follows. First solve the shortest-path problem for the N-node network for all pairs of initial and final nodes. Let d'_{ij} represent the length of the shortest path from node i to j. Then, solve the $(k+1)$-city 'traveling-salesman' problem for the shortest path from 1 to N passing through nodes 2, 3, $\cdots$, k, where the distance from node i to j is d'_{ij}. Reference 35 discusses methods of solution. While no easy solutions exist for the traveling-salesman problem, the specified-city problem can certainly be no easier than the traveling-salesman problem of dimension $k+1$, since if $k=N-1$ it *is* the traveling-salesman problem.

6. CONCLUSION

WE HAVE referenced, evaluated, and occasionally modified various algorithms for computationally solving certain shortest-path problems. By collecting contributions from several disciplines, we hope to re-orient and revitalize the sometimes rather incestuous research of each. The author would appreciate notification of overlooked or new significant research in this area.

NOTES

1. M. Bellmore points out, in a private communication, that if arcs of length 0 are permitted, special care must be taken to prevent possible cycling during the phase of the calculation used to deduce the path from the value table.

2. In giving this credit to Minty, Pollack and Wiebenson cite, without date, a private communication from Minty.

3. The additional computation introduced in order to test for the presence of infinities adds about 5 per cent to the computing time. For a sample problem with 10 nodes and 34 arcs and, hence, 56 infinities initially, the test yielded a 5 per cent net improvement over no test. The amount of net improvement or degradation depends on the actual network configuration as well as the number of nonexistent arcs.

ACKNOWLEDGMENT

THIS RESEARCH was supported by the US Air Force under Project Rand and by a National Science Foundation grant to the University of California.

REFERENCES

1. A. H. LAND AND S. W. STAIRS, "The Extension of the Cascade Algorithm to Large Graphs," *Management Sci.-A* **14,** 29–33 (1967).

2. T. C. HU, "A Decomposition Algorithm for Shortest Paths in a Network," *Opns. Res.* **16,** 91–102 (1968).

3. ——, *A Short Cut in the Decomposition Algorithm for Shortest Paths in a Network,* Mathematics Research Center, US Army, Report No. 882, The University of Wisconsin, Madison Wisconsin, July 1968.

4. E. W. DIJKSTRA, "A Note on Two Problems in Connexion with Graphs," *Numerische Mathematik* **1,** 269–271 (1959).

5. P. D. WHITING AND J. A. HILLIER, "A Method for Finding the Shortest Route through a Road Network," *Opnal Res. Quart.* **11,** 37–40 (1960).

6. M. POLLACK AND W. WIEBENSON, "Solution of the Shortest-Route Problem—A Review," *Opns. Res.* **8,** 224–230 (1960).

7. L. R. FORD, JR. AND D. R. FULKERSON, "Constructing Maximal Dynamic Flows from Static Flows," *Opns. Res.* **6,** 419–433 (1958).

8. G. B. DANTZIG, "On the Shortest Route through a Network," *Management Sci.* **6,** 187–190 (1960). *See also* Ref. 10.

9. C. BERGE AND A. GHOUILA-HOURI, *Programming, Games and Transportation Networks,* p. 176, trans. by M. MERRINGTON AND C. RAMANUJACHARYULU, Methuen, London, 1965.

10. G. B. Dantzig, *Linear Programming and Extensions*, pp. 363–366, Princeton University Press, Princeton, New Jersey, 1963.

11. T. A. J. Nicholson, "Finding the Shortest Route Between Two Points in a Network," *Computer J.* **9,** 275–280 (1966).

12. J. D. Murchland, *The "Once-Through" Method of Finding All Shortest Distances in a Graph from a Single Origin*, Transport Network Theory Unit, London Graduate School of Business Studies, Report LBS-TNT-56, August 1967.

13. L. R. Ford, Jr., *Network Flow Theory*, The Rand Corporation, P-923, August 1956.

14. E. F. Moore, "The Shortest Path through a Maze," pp. 285–292, *Proc. Int. Symp. on the Theory of Switching*, Part II, April 2–5, 1957, The Annals of the Computation Laboratory of Harvard University 30, Harvard University Press, 1959.

15. R. E. Bellman, "On a Routing Problem," *Quart. Appl. Math.* **16,** 87–90 (1958)

16. J. Y. Yen, *Matrix Algorithm for Solving All Shortest Routes from a Fixed Origin in the General Networks*, presented at the Second International Conference on Computing Methods in Optimization Problems, Sept. 9–13, 1968, San Remo, Italy.

17. G. B. Dantzig, W. O. Blattner, and M. R. Rao, *All Shortest Routes from a Fixed Origin in a Graph*, Operations Research House, Stanford University, Technical Report 66-2, November 1966; also in *Théorie des Graphes*, pp. 85–90, Proceedings of the International Symposium, Rome, Italy, July 1966, published by Dunod, Paris.

18. L. E. Hitchner, *A Comparative Investigation of the Computational Efficiency of Shortest Path Algorithms*, Operations Research Center, University of California, Berkeley, Report ORC 68-17, July 1968.

19. C. Witzgall, *On Labelling Algorithms for Determining Shortest Paths in Networks*, National Bureau of Standards Report 9840, US Department of Commerce, Washington, D. C.

20. R. W. Floyd, "Algorithm 97, Shortest Path," *Comm. ACM* **5,** 345 (1962).

21. S. Warshall, "A Theorem on Boolean Matrices," *J. ACM* **9,** 11–12 (1962).

22. J. D. Murchland, *A New Method for Finding All Elementary Paths in a Complete Directed Graph*, Transport Network Theory Unit, London School of Economics, Report LSE-TNT-22, October 1965.

23. G. B. Dantzig, *All Shortest Routes in a Graph*, Operations Research House, Stanford University, Technical Report 66-3, November 1966; also in *Théorie des Graphes*, pp. 91–92 Proceedings of the International Symposium, Rome, Italy, July 1966, published by Dunod, Paris.

24. A. Shimbal, "Structure in Communication Nets," *Proc. of the Symposium on Information Networks*, Polytechnic Institute of Brooklyn, April 12–14, 1954.

25. "Investigation of Model Techniques," *Second Annual Report, July 1957–1958*, Case Institute of Technology, Cleveland, Ohio, ASTIA Report AD211968.

26. T. C. Hu, "Revised Matrix Algorithms for Shortest Paths," *SIAM J. on Appl. Math.* **15,** 207–218 (1967).

27. W. Hoffman and R. Pavley, "A Method for the Solution of the Nth Best Path Problem," *J. ACM* **6,** 506–514 (1959).

28. R. Bellman and R. Kalaba, "On *k*th Best Policies," *J. SIAM* **8,** 582–588 (1960).

29. M. Pollack, "Solutions of the *k*th Best Route through a Network—A Review," *J. Math. Anal. and Appl.* **3,** 547–559 (1961).

30. M. Sakarovitch, *The k Shortest Routes and the k Shortest Chains in a Graph,* Operations Research Center, University of California, Berkeley, Report ORC 66–32, October 1966.

31. S. Clarke, A. Krikorian, and J. Rausen, "Computing the *N* Best Loopless Paths in a Network," *J. SIAM* **11,** 1096–1102 (1963).

32. M. Pollack, "The *k*th Best Route through a Network," *Opns. Res.* **9,** 578–580 (1961).

33. K. L. Cooke and E. Halsey, "The Shortest Route through a Network with Time-Dependent Internodal Transit Times," *J Math. Anal. and Appl.* **14,** 493–498 (1966).

34. J. P. Saksena and S. Kumar, "The Routing Problem with 'K' Specified Nodes," *Opns. Res.* **14,** 909–913 (1966).

35. M. Bellmore and G. L. Nemhauser, "The Traveling Salesman Problem: A Survey," *Opns. Res.* **16,** 538–558 (1968).

POSTSCRIPT

Following is a recent result, which underlines an important distinction between *adaptive* and *non-adaptive* algorithms that did not receive sufficient emphasis in the survey paper.

In describing the Dijkstra algorithm I assumed a very naive method of distinguishing "labelled" from "unlabelled" nodes which required N^2 comparisons. Yen has since observed that this can be accomplished in $0(N)$ comparisons, so the Dijkstra procedure solves the initial-node-to-all-other-nodes problem in $N^2/2$ additions and N^2 comparisons plus calculations of order N.

The implication of this is that the all-pairs problem can be solved by N applications of Dijkstra's algorithm in $\frac{N^3}{2}$ additions and N^3 comparisons, which is better than Floyd's algorithm. Empirical evidence verifies Dijkstra's superiority. This is surprising until one realizes that the Dijkstra algorithm is adaptive (which node it processes next depends on the current problem data) while Floyd's algorithm processes nodes in the same order regardless of data. While Floyd's scheme is thought to be the best possible non-adaptive procedure, no one has any idea about lower bounds for adaptive procedures.